Migration in Social Ecology of North East Himalayas and Visualization through Mental Modeling

About the Author

Prof. (Dr.) Sankar Kr Acharya, presently, **Dean, Post Graduate Studies; f**ormer Head, Department of Agril Extension and Director, Extension Education, Bidhan Chandra Krishi Viswavidyalaya, Mohanpur, WB, born on 6th October, 1960, started his career as Assistant Professor at BCKV in 1988 and has been in teaching, research and extension over 34 years. An erudite teacher as well as an elegant speaker, is internationally acclaimed for his unique research domain of Social Entropy and Energy Metabolism, Social Ecology and Environmental Sociology, Enterprise Ecology Framework, Farm Stewardships: The Transition from Conservation, Technology Socialization Process, and The Social Ecology of Livelihood, which have got tremendous policy as well as scholastic importance. Research Publication: 239 in National and International Journals. Book Publication: 110 books authored / co-authored. Expert Member, Agriculture Commission; Marketing and Extension Sub Committee, Govt. of West Bengal; Expert member WWF projects in BTR (Buxa Tiger Reserve Project); Expert member; DFID project on Primary Education (DPEP); Rapid Environment Impact Analysis, Visited Italy, France, Germany, China, Sri Lanka and Bangladesh. Delivered 31 Key note addresses, 13, invited and lead speakers in International Congress, Chaired 19 Sessions in International conferences, and Editors of a score of national and international journals. He has so far been awarded with 15 best paper awards in National and International Conferences Organizing Secretary of two International Conferences, held in the years, 2016 & 2018, Jointly Organized by Department of Agricultural Extension, BCKV, WB and Krishisanskriti, New Delhi. He has also been honoured to be selected as the Convener of Panel (PE-32) entitled The hunger, poverty and silence, of IUAES, University of Manchester, UK, 2013 .Co-PI of ICAR, NAHEP and World Bank funded project on Conservation Agriculture. So, far he has successfully guided 19 Ph D scholars, of which one awarded with DAAD fellowship, Germany and two other placed as visiting scholars at Ohio State University and Cornel University, 2020-21, USA; 83 M Sc students from BCKV CU, KU, Vidyasagar University. He is the fellow of ISEE, IARI, New Delhi, Fellow, BIOVED, He is now acting as editor, reviewer of a good number of national and international journals. Fellow Award, ISEE(ICAR-IARI, New Delhi), 2019 Awards and Distinctions: Distinguish Scholar Award, Krishisanskriti, New Delhi; Certificate of Merit for standing First class first at M Sc(Ag) in Agricultural Extension, 1986; Honorary Appointment to the Research Board of Advisors, The American Biographic Institute (2003); He has been honored to be selected as the Convener of Panel (PE-32) entitled the hunger, poverty and silence, of World Congress, IUAES, University of Manchester, UK, 2013 ;He has been honored to be selected as the Convener of Panel (P-102) entitled, Uncertainties, Unpredictability and Marginalization: The impact and mitigation for survival of agriculture and humanity, of the 19th World Congress, IUAES-WAU, 14-20 October, 2023, New Delhi, India, 2023 ;Honored to be nominated as Expert Member, Working group on Agricultural Extension and Training, West Bengal State Agriculture Commission (2007); Honored to be nominated by the Hon'ble Vice-Chancellor, BCKV to Act as an Expert Member, DPEP, A DFID, UK, project for planning and revamping primary education through POA of GOVT of India.(1995) Expert Member, Buxa Tiger Reserve Project funded by WWF(World Wide Wild Life Funding), 1998; Distinguished Professor and Academician Award, Research Education Solution (RES), M S Swaminathan School of Agriculture, Centurion University(2022); Diamond Achievers Award, Council for Academic Performance and Appraisal, New Delhi(2022); Honorary Fellowship Award, BIOVED Research Institute of Agriculture and Technology, Prayagraj on Occasion of Agricultural Scientists and Farmers' Congress. Achievers of German patent (Utility model no. 20202310273): Novel LOT based inventory management system with radio controlled pallet racking for storage solution.

Mr. Sarup Chettri, born on July 20, 1998, in the picturesque town of Darjeeling, is a dedicated scholar in the field of agriculture. He earned his B.Sc. Agriculture Hons degree from Bidhan Chandra Krishi Viswavidyalaya, showcasing a profound commitment to the agricultural sciences. Furthering his academic journey, Sarup pursued an M.Sc. in Agriculture with a specialization in Agriculture Extension Education. His research expertise came to the forefront through a thesis centered on the "Application of Mental Modeling in assessing and delineating the Social Ecology of Migration in Darjeeling." This work not only demonstrated his academic prowess but also showcased a keen interest in understanding the intricate dynamics of social ecology in his hometown. He had his M Sc.(Ag.) dissertation guided by Prof S K Acharya, the eminent professor of Extension Science in India. Sarup Chettri's academic achievements and research focus highlight his passion for contributing to the agricultural and social development of the Darjeeling region.

Mr. Arnab Banerjee, born on 25thFebruary 1997, in Bankura district of West Bengal, obtained his B.Sc. (Agriculture) Honours from Bidhan Chandra Krishi Viswavidyalaya, Mohanpur, West Bengal, in the year 2019. He obtained his M.Sc. in Agriculture and Rural Development from Ramakrishna Mission Vivekananda Educational and Research Institute, Narendrapur, West Bengal, in the year 2021. He has published some research articles in reputable journals. Additionally, he has been awarded with the best oral presenter awards in international conferences. Currently, he is pursuing his Ph.D. in the Department of Agricultural Extension at the Bidhan Chandra Krishi Viswavidyalaya under the supervision of Prof. S.K. Acharya.

Mr. Monirul Haque, born on 9th December 1995, in Falakata of Jalpaiguri district, which is presently in Alipurduar district of West Bengal, obtained his B.Sc. (Agriculture) Honours from Palli Siksha Bhavana (Institute of Agriculture), Visva-Bharati, Sriniketan, West Bengal, in the year 2017. He obtained his M.Sc. (Ag.) in Agricultural Extension in the Department of Agricultural Extension from Bidhan Chandra Krishi Viswavidyalaya, Mohanpur, West Bengal, in the year 2019. He qualified the National Eligibility Test for Assistant Professor conducted by UGC and ICAR-ASRB in his very first attempt in the year 2019. Additionally, he has published several research articles in reputable journals. Currently, he is pursuing his Ph.D. in the Department of Agricultural Extension at the same university under the supervision of Prof. S.K. Acharya. He has been awarded the Maulana Azad National Fellowship for Minority Students to support his doctoral degree program.

Migration in Social Ecology of North East Himalayas and Visualization through Mental Modeling

Sankar Kr Acharya
Sarup Chettri
Arnab Banerjee
Monirul Haque

FU 75, Pitam Pura
New Delhi 110034
publishmywork23@gmail.com
Mob. 91-8447075807

Print ISBN: 978-81-96458-87-4

ebook ISBN: 978-81-96458-88-1

Preface

Migration is an inevitable character of evolving civilization. It was there, it is now everywhere and it will continue for the posterity. Two factors are responsible for migration i.e. Push factors and Pull factors. Due to push factors the people under compulsion migrate from their home and on the other hand the pull factors attracts towards outside location and system because of their advantageous socio-economic and ecological situation and people leave their home so one is under compulsion and the other is through choice. Darjeeling region is internationally acclaimed and known as one of the most beautiful hill station and tourist destination in the world because of its scenic landscape, mesmerizing tea gardens, flora and fauna and local hospitality but right at this moment it has been observed that the young populace from Darjeeling are migrating to other parts of India and this has been due to the decline in tea productivity, deterioration in surrounding agricultural economy, socio-political unrest, degeneration in tourism economy so the present study focused its attention on these migration aspect with the title " Application of Mental Modeling in assessing and delineating the Social Ecology of Migration in Darjeeling." A score of 80 respondents have been selected from Happy Valley Tea Garden and Tukver Tea Garden of Darjeeling-Pulbazar CD Block and PulungdungKhasmahal and PokhriabongKhasmahal of Jorebunglow-SukhiaPokhri CD Block. A plethora of statistical tools has been applied to elicit the empirical facts which are responsible for migration. It has been found that variables like age, family income, social support network, duration of stay at home, frequency of visiting home, land holding, family size, no. of information sources and distance of migration have come out as the marker variables and contributed substantially in the form of functional impact on migration. Mental Modeling; Fuzzy-logic Cognitive Mapping (FCM) has been applied to elicit the migrant's perception, assumption and decision-making and after subsequent examination of different scenario analysis it has been found that Scenario I reveals that when components like poverty, insecurity and risk were reduced and livelihood when enhanced, job opportunities, social network, lifestyle, status, and economic satisfaction were increased and migration showed a decline. Scenario II reveals that job opportunities, educational opportunities, and entrepreneurial opportunities when enhanced, income showed a noticeable increase and migration declined significantly. Additionally healthcare facilities also showed an increase. So the study presents a unique combination of multivariate analytical techniques and participatory mental modelling. The empirical study suggests some of the vital recommendations which could have restrained the young population of Darjeeling from migration and these are improvement in job opportunities, restoration of socio-political stability and its maintenance, augmentation of tourism activities and its economy, restoration of tea garden economy and livelihood generation process, restoration of ecological resilience across the terrains of Darjeeling and surrounding areas of Kalimpong. Unless the systemic potentials of Darjeeling is being restored, upgraded and evolved, the social equilibrium of Darjeeling will be disrupted. This is equally detrimental for the economy, ecology and sociology of Darjeeling, one of the most beautiful hill-station in the world.

The book, Migration in Social Ecology of North East Himalayas and Visualization through Mental Modeling, as I trust and believe, shall go a long way in contributing policy research and empirical studies in regard to migration and its impact in any operation social ecology across the globe. It is going to be a worth reading document for all scholars and faculties hailing from multidisciplinary background covering international audience.

Authors

Contents

Preface .... *vii*

1. Prologue 1
2. Review and Citations 7
3. Theory and Concepts 17
4. Social Ecology 49
5. Methods and Material 77
6. The Case Studies and Visualization 93
7. Epilogue 123
8. Limitations and Future Scope 127

Bibliography *129*

1

Prologue

Migration is a worldwide occurrence that can be attributed to various factors such as disparities in economic development, regional differences, and varying living standards among socio-economic groups. It is driven by both push and pull factors. Pull factors include opportunities for better employment and higher wages, while push factors encompass the lack of job prospects in underdeveloped areas, droughts, and scarcity conditions. Unlike the biological factors of mortality and fertility, which operate within a biological framework, migration is influenced by the desires of individuals. Although most migratory movements are intentional, there may be exceptional cases where this is not true. Therefore, migration is a response by individuals to economic, social, political, and demographic forces in their environment and serves as an important indicator of social change in society.

The term "migration" encompasses various meanings and interpretations, depending on the nature, scope, or purpose of the study or discussion. Sociologists highlight the social and cultural consequences of migration, while geographers emphasize the significance of time and distance. Economists, on the other hand, give importance to the economic aspect of migration. Dictionaries generally define migration as a change in residence from one place to another. The **Oxford Advanced Learner's Dictionary** of current English defines migration as "moving from one place to another (to live there)," while the **Webster's New World Dictionary** states that it refers to "moving from one place to another, especially to another country" or "moving from place to place to harvest seasonal crops." The **Demographic Dictionary** describes migration as a form of geographical or spatial mobility involving a change in residence from the place of origin or departure to the destination or arrival. According to the **Inter-State Migrant Workmen (Regulation of Employment and Conditions of Service) Act of 1979**, a migrant workman is defined as an individual who is hired through a contractor in one state under an agreement or arrangement to work in an establishment located in another state. This definition applies regardless of whether the principal employer of that establishment is aware of the recruitment or not.

Migration is a complex phenomenon that has influenced societies and human history for many years. In search of better prospects, safety, or a different way of life, it involves the movement of individuals or groups of people from one place to another, frequently across national or international lines. The causes of migration are numerous and varied, ranging from monetary factors and political unrest to alterations in the environment and social dynamics. Migration, along with fertility and mortality, is a crucial indicator of shifts in the social and economic domains. It encompasses the movement of people between different regions within a country and across international borders. The historical roots of both types of migration in India can be traced back to the nineteenth century, if not even earlier (Chattopadhyaya, 1987).

Migration is a field of research that allows for a multidisciplinary approach within the social sciences. It encompasses various disciplines such as social demography, sociology, social and cultural anthropology, economics, history, geography, and psychology (Rao, 1986). The term "migration" refers to the movement of a population involving a change in residence. This change can occur between villages or towns, from a town to a village, within a district or commune, between districts or communes, within a country, or even outside the country. The duration of the migration can range from one month to several months or more.

It is recognized that the main driving force behind migration is the pursuit of better job prospects, and the extent and direction of migration are predominantly influenced by employment opportunities (Khan, 1983). Therefore, the migration history can be viewed as a series of shifts in labour reservoirs (Bolaria, 1997).

Understanding migration necessitates a thorough investigation that looks past personal motives and into the larger social, economic, and environmental settings in which migration takes place. The social ecology of migration is one theoretical paradigm that provides such a comprehensive viewpoint. The discipline of social ecology is concerned with the dynamic interactions between people, society, and their environments. Social ecology is a field of research that examines how social, economic, and environmental factors affect the patterns, processes, and results of migration.

Migration arises as a result of changes in the economic and political landscape. Economic development leads to inequalities in wealth, both between countries and within different regions of a country. These disparities act as incentives for individuals to move from areas with limited opportunities to regions that offer higher prospects. It is evident that the majority of migrants, whether they are crossing international borders or moving within a country, experience economic benefits from their relocations (Walter, 2002).

The social ecology of migration is based on the understanding that migration is a social phenomenon' entangled in a complex web of connections, structures, and systems. It recognizes that migration is influenced by broader social influences, such as family networks, labour markets, political institutions, and cultural norms in addition to being a decision made by an individual. These social variables combine with environmental variables like re*Source* shortages, natural catastrophes, and climate change to produce complicated push and pull variables that affect migration movements.

Additionally, different migrant populations' unique experiences and vulnerabilities are acknowledged by the social ecology of migration. It looks at how racial, ethnic, gendered, class, and legal status differences interact with social and environmental dynamics to influence migrant experiences. For instance, institutional discrimination and inequities frequently present additional difficulties and impediments for marginalised groups like refugees, undocumented migrants, or indigenous populations.

Furthermore, the social ecology of migration emphasises the impact of community dynamics and social networks on migration trends. For navigating the migration process and settling into new contexts, migrants frequently rely on official and informal social contacts. These networks offer re*Source*s, assistance, and information that can help or hurt the relocation process. Understanding migrants' social networks and community dynamics is essential to comprehending their integration, wellbeing, and potential effects on both sending and receiving societies of migration.

The interconnectedness between sending and receiving groups is further highlighted by the social ecology of migration. It acknowledges that migration is not a one-way process but instead involves a mutually beneficial link between the origin and destination regions. The communities that people move to and from have an impact on both the communities they leave behind and the communities they settle in. This is true for both individuals and families that migrate. It may have important economic, social, and cultural repercussions for both sending and receiving nations, which call for analysis within a thorough social ecological framework.

Migration can serve as an important means to facilitate economic development to some extent. This is because migration allows for the movement of human re*Source*s from areas where their social contribution is assumed to be minimal or negligible to places where their productivity can significantly increase due to capital accumulation and technological advancements. Simultaneously, migration plays a crucial role in fulfilling the demand for various types of manpower in different locations. As described by Todaro, the migration of workers can be seen as a socially and economically beneficial process, as it enables individuals to move from areas with low productivity and labour shortages. Baque suggests that migration is influenced by both "push" and "pull" factors at the place of origin, where migration occurs when the positive factors attracting people to the destination outweigh the negative factors pushing them away from their place of origin. Myrdal emphasizes push factors as the main drivers of migration, while Trewartha, Sovani, and Bas explain that migration results from the complex interplay between push and pull factors at the origin.

Migration is also influenced by factors such as the geographical and locational characteristics of migrants' places of residence (Mehta, 1991). As a subject within population studies, migration is primarily focused on empirical analysis. There is still a significant emphasis on data measurement and provision, potentially overshadowing the development of theoretical frameworks. Nonetheless, migration is an integral part of development and has received considerable attention in the literature on development. Everett Lee's approach involved a descriptive model of migration, considering a series of "push" factors from areas of origin and "pull" factors toward destination areas. The factors that drive migration are commonly attributed to two main forces: population growth in rural areas, which exerts pressure on agricultural re*Source*s and compels people to leave, and economic conditions primarily influenced by external factors that attract people to urban centres (Skeldon, 1997).

The significance of migration for developing nations cannot be overstated. It plays a crucial role in economic development and manpower planning. In the context of the commercialization of agriculture, migration has gained special importance due to labour mobility. It is a major driver of urbanization and brings about social change. It is important to view migration and settlement as interconnected aspects of the social and cultural lives of people. Migration involves a change in the place of residence for a certain period of time, excluding short visits and tours, and encompasses various forms of both voluntary and involuntary movements (Rao, 1986). There are also situations where migration is essential for people to earn a livelihood, such as nomads, shifting cultivators, itinerant traders and salesmen, artisans, and labourers.

The phenomenon of migration can be comprehended within the framework of social relationships and is inherently linked to human social interactions. Therefore, in order to engage in a comprehensive discussion on migration, it is crucial to examine the population structure of a given region, including factors such as population size, population growth, and various demographic characteristics. These characteristics encompass socio-cultural, socio-economic, and migration-related aspects of the individuals residing in that particular region (Datta, 2003).

1.1 Research Questions

The broad research questions those are attempted to explore in the study are:

1. What are the reasons behind people's decision to move from one place to another?
2. What are the characteristics and trends observed in migration?
3. How can we explain the functional relationships between different variables being studied, both within and between different levels of interaction?
4. How do participatory tools and techniques assist in uncovering the true reality and visualizing the features and factors being studied?

5. To what extent do empirical evidence and concrete data contribute to the formulation of policies within the micro-sociological context being examined?

1.2 Need for the Study

1. Relatively few attempts have been made to identify the different factors or variables responsible for migration and relationship between them in the Darjeeling region.
2. To focus on the socio-economic attributes of the migrant's households in the study area.
3. To identify the push and pull factors of migration as characterizing the level, its change and different phases of migration.
4. To make an analysis of determining the differentials and determinants of migration process in the study area which will be of immense help for policy makers and rural development programme initiators to develop a policy implication for an overall development in the area.

1.3 Objectives of the study

Based on the above research questions, the broad objective of this study is to understand the complex nature of family migration decision and the factors those are responsible for this migration decision in the study area. The specific objectives of the present study are as follows:

1. To delineate the present status of Migration with the perspective of Social ecology of Darjeeling.
2. Application of Mental Modeling to elicit the binary of choices in different forms of Migration in Darjeeling.
3. To elucidate the intra and inter level interaction amongst and between the sets of variables,viz., Typology & Dynamics of migration as consequent variable and its score of socio-ecological and economic variables.
4. To operate a micro-level policy implication based on the empirical study.

2

Review and Citations

A thorough examination of relevant literature is crucial in any scientific investigation and research. Apart from identifying previous work conducted in the problem area or the specific field of study, the primary function of a literature review is to provide insights into the methods and procedures used, while also establishing a foundation for interpreting the findings. A literature review involves a comprehensive and critical assessment of prior research, offering a summary and overview of a particular research area. This allows readers to understand why the current research program is being pursued. A well-executed literature review delves into the reasons behind selecting a specific research question, integrating and evaluating previous studies, and demonstrating how they relate to the proposed research program. It is important to present all sides of the argument in a clear manner to avoid bias, while also highlighting areas of agreement and disagreement among the existing literature.

In this chapter, we have attempted to provide a concise overview of various studies and writings related to migration. This overview serves to enhance our understanding of the significance and purpose of our current research. Some relevant reviews have been categorized and presented under the following sub-headings:

2.1 General Survey on Migration

Collinson, S et al, (2009) in his study concluded that Migration patterns and dynamics are incredibly diverse, intricate, and context-specific. Investigating the interactions between local-level factors that directly affect people's migration decisions and strategies (linked to livelihoods) and interconnected political, economic, and social factors and processes at various levels that affect the agency, vulnerability, and/or opportunities of migrants (and associated non-migrants) during the migration process and that shape migration outcomes may help us better understand these dynamics in particular contexts. Since these are essential to comprehending the interests and relative agency of individuals, households, and other actors involved, different and shifting power relations are crucial to understanding the interaction of local migration strategies with political and economic processes and relationships affecting these strategies.

McCarty et al, (2006) suggests that the agricultural sectors in rural areas of transitional nations have undergone structural upheaval over the past few years, along with significant demographic shifts brought on primarily by widespread emigration to metropolitan areas and abroad. Very little is known about the connections between these activities, despite the potential importance of migration and the remittances that follow in supporting or impeding changes in agriculture. Findings imply that migration significantly reduces agricultural labour per capita. The findings does, however, also imply that the loss of household labour in agriculture is offset by improved access to capital, resulting in increases in both agricultural and general incomes.

The expense of transportation is a significant element that affects migration patterns. Research examining regional data reveals that individuals are more inclined to migrate shorter distances rather than longer distances. This preference is not only due to lower costs but also because it is easier for them to gather information about the culture, language, and opportunities available in nearby locations. Studies conducted by Richey (1976) and Gardner (1981) on both developed and developing countries have discovered that this tendency for short-distance migration is particularly prevalent among low-income and less-educated migrant groups.

Schultz's (1971) research demonstrated that the distance of migration, which serves as an indicator of migration costs, can sometimes act as a barrier to certain types of migration. International migration, for instance, involves substantial expenses. According to Schultz, individuals who migrate internationally are generally more educated, earn higher incomes, possess greater risk tolerance, and have more valuable job skills compared to internal migrants.

Kuznets (1966) provided an explanation of how economic growth and development can contribute to migration. As technological advancements occur, the demand for workers in various industries undergoes changes. In agriculture, for instance, technological improvements lead to a reduced need for labor. According to Kuznets, migration should serve as a means to compensate for the faster growth in labor demand outside the agricultural sector, as well as to accommodate the higher rate of natural increase in agricultural labor. This is because population growth in rural areas tends to be higher than in urban areas.

The Todaro (1969) model, a groundbreaking contribution to migration studies, is primarily focused on the economic aspects of rural-urban migration. It builds upon Lewis's (1954) initial model of development and the extended framework developed by Fei and Ranis (1961, 1964). The Todaro model highlights that migrants consider various labor market opportunities between rural and urban sectors and select the option that offers the highest expected gains from migration. The model operates under the following assumptions:

a) Both current and potential members of the workforce compare their anticipated incomes in the urban sector over a specific timeframe with the existing average incomes in rural areas. If the former exceeds the latter, individuals decide to migrate.

b) The decision to migrate is based on the expected wage differentials between urban and rural areas, rather than the actual differentials. The "expected" differential is determined by the interaction of two factors: the real wage difference between urban and rural areas and the probability of successfully securing employment in the modern urban sector.

c) The likelihood of obtaining an urban job is inversely related to the level of urban unemployment.

Furthermore, the model provides insights into the functioning of urban labour markets and how immigrants are assimilated into those markets.

Various researchers have conducted tests on the fundamental Todaro migration model, yielding different results. For instance, Barnum and Sabot (1977) examined data from Tanzania and found supportive evidence for the model. Similarly, Yap (1977) conducted a study in Brazil, which indicated that migrants make rational decisions based on earnings differentials. However, Yap's findings contradicted Todaro's observation regarding the duration of job searches for migrants. Yap discovered that migrants had shorter job search durations but earned less than non-migrants. He attributed this difference to factors such as age and skills, rather than the migrants' status. Additionally, Papola (1981) observed low unemployment rates and high labour participation rates among migrants. He also noted that city in-migrants, particularly those without skills, had relatively short job search durations. However, Papola did not support Todaro's prediction that migrants would earn less than non-migrants.

McCatty (2004) concentrates on the phenomenon of rural-urban migration and its impact on urbanization, particularly in developing nations. In his study, he examines three theoretical models that explain the process of rural-urban migration: the Lewis dual sector model, the family/household migration model, and the Todaro model. He argues that rural-urban migration is an inevitable outcome influenced by both asymmetric policies and the economic development of urban areas. Consequently, migration should be seen as a response that brings equilibrium to the existing economic disequilibrium. As long as there are disparities in employment opportunities between rural and urban areas due to urban bias, migration will persist. It is the government's responsibility to mitigate this disequilibrium.The paper also assesses the benefits and costs associated with rural-urban migration. While there are potential benefits that can arise from such migration, the costs are widespread in developing countries. The marginal social costs outweigh the marginal private costs, making urbanization inefficient. The level of urbanization reaches a point where the marginal private costs are equal to the marginal private benefits.

The influence of remittances on migration has been examined by Glytsos (1988). He considers remittances as a significant and internal factor in the decision-making process of temporary migration within families. The model is primarily tested using data from Greek-West German temporary migration and has shown strong performance. Its validity is further supported by an additional test using data from Greek American permanent migration. Funkhouser (1995) observed the determinants of remittances by analysing household data from El Salvador and Nicaragua. In some cases, out-migration occurs as a result of debt back home or debt-interlocking arrangements, where debts are repaid through labour commitments or "dadan" involving employers or intermediaries in the destination areas. This type of migration may or may not address the underlying causes of the debt. Additionally, reducing personalized dependencies or interlocked relationships can also contribute to increased labour mobility and migration, as individuals seek alternative Sources of cash income. (Srivastava, 1987; Breman, 1974, 1985; Mosse et al., 1997).

Numerous studies have revealed that demographic factors play a significant role in migration. Schultz (1971) conducted a study using Colombian Census data and discovered that young individuals migrate at a rate proportional to population growth. According to his findings, a 10 per cent difference in the population growth rate led to a 25 per cent difference in the out-migration rate among those under the age of 26. Gugler (1968) conducted research on Africa and observed that 42 per cent of migrants belonged to the age group between 10 and 39. Additionally, family size, structure, and any conflicts within the family unit are important considerations in the decision-making process of migration.

Hossain (2001) conducted a study in ten villages of Cumilla district, Bangladesh, focusing on rural-urban migration. The study aimed to examine the differences and factors influencing migration patterns. The findings revealed that adults with higher education levels were more likely to engage in rural out-migration. Prior to migrating, a majority of the individuals were either studying or unemployed. Approximately half of the migrants moved for temporary employment, while about a quarter sought permanent jobs. Moreover, the study established a correlation between educational attainment and permanent migration, with illiterate migrants predominantly engaging in temporary migration. The migration rate was significantly higher among the educated and unemployed individuals, particularly within the age group of 20-29.According to Hossain, the main factors attracting individuals to migrate to urban areas were better opportunities, the presence of previous migrants, and job availability. Conversely, factors such as poverty, job search difficulties, and family influence were identified as primary drivers for rural out-migration.

In many cases, women participate in migration alongside male members of their households. It is common for younger siblings and older children to accompany their parents and work alongside them. When a family migrates, it often involves the younger members of the family, leaving the elderly to handle additional responsibilities while also meeting their own subsistence and basic needs **(Mosse*et*** al., 1997 and 2002). The impact of male migration can be particularly challenging for girls, as they often have to assume additional domestic responsibilities and care for younger siblings. The absence of male supervision further diminishes their opportunities for education (Srivastava, 2001).

Scrasee t al, (2016) points out that major change have occurred in Darjeeling as a result of greater rural mobility, rapid rural-to-urban migration, and increased rural engagement in municipal governance. We show that the formerly inferior rural classes' upward trajectory is making more affluent urban inhabitants feel threatened, which causes local political issues to be redrawn along rural-urban lines and a reconfiguration of class consciousness and social interactions. The urban middle class, whose opportunities in the city have stagnated or decreased views rural migrants as a Source of competition for reSources that are becoming increasingly limited and holds them responsible for the general decline in the quality of urban life. They use their (mostly cultural) capital to reinforce cultural markers that set them apart from rural migrants.

Curran et al, (2006) points out that the fundamental ways in which gender fundamentally organizes the social relations and institutions determining the causes and consequences of migration are evident from a survey of sociological research on gender and migration. Yet, despite the fact that gender and migration have been the subject of major sociological research over the past three decades, these studies are not equally dispersed among the field.

Acharya et al, (2022) considered in his study that push and pull factors of rural migration, especially from agrarian ecology to elsewhere, may be into an urban social ecology by using comprehensive multivariate statistical design analyses and design, it has isolated a score of marker variables, responsible for characterizing and organizing rural-urban migration of people from a marginalizing economic and social echelon.

deHann (2011) has posed an important question that directly addresses a specific gap in migration literature. The question pertains to whether studies on migration have become an essential component in comprehending development processes such as economic growth, poverty reduction, and rural development. The argument presented suggests that these processes do not necessarily result in a reduction of migration. However, despite the need for evidence to support this claim, it remains challenging to incorporate a comprehensive understanding of mobility, particularly labour migration, into mainstream texts on development studies.

2.2 Migration Decision Making

There are two types of migration decisions: individual decision making and family decision making. Each of these decision-making processes involves different factors. As Stark (1991) accurately highlighted, the family should not be treated as if it were an individual. Therefore, there are differences between individual migration decisions and family migration decisions. The decision to migrate is influenced by a variety of factors, including economic, demographic, and sociological factors. However, individual and family migration decisions are shaped by distinctly different sets of factors.

2.2.1 Individual Migration Decision

Individual migration decisions have been referred to as the "place utility" of specific migrants by Wolpert (1965). He argued that migrants select particular destinations based on the level of satisfaction or "highest place utility" they offer. This concept of utility does not solely revolve around expected wages, as proposed by Todaro (1969).Other factors, such as psychological considerations, also play a role, compelling migrants to choose locations that are closer to their relatives or native places. To further explore the reasons behind migration, the study introduced the "stress-threshold" model, which emphasizes the tolerance of a certain level of discomfort or "stress" in the migrants' place of origin. The study assumes that individuals have a threshold level that determines whether they choose to migrate and where they migrate to. Ultimately, migrants select destinations that provide the highest level of place utility, which, according to the study, represents the actual process of "choosing a destination."

The value-expectancy model, another branch of the behavioural model of migration as identified by Crawford (1973), suggests that migrants make deliberate decisions to migrate based on factors beyond mere economic considerations. This study highlights that the likelihood of migration is determined by multiplying the values individuals attach to potential migration outcomes with their expectations that migration will indeed lead to these outcomes. These values and expectations are influenced by personal and household characteristics, such as education level, as well as societal norms.In addition, the utility of migration literature itself appears to encompass a broader decision-making framework for individuals, as discussed by Harbison (1981). This framework emphasizes that the decisions of individual migrants are significantly influenced by their families through the demographic structure of the latter. Factors such as the number of dependents in the family and the economic status of the family can particularly impact the migration decisions of women. Aside from economic motives, there are also social constraints that compel women to migrate with their husbands, even if they do not personally benefit from the migration experience.

Lee (1966) was the pioneer in developing a push-pull framework for understanding migration at the individual level. This framework takes into account factors from both the supply and demand sides of migration. The model incorporates various positive and negative factors, both at the origin and destination, which either push individuals to migrate or pull them towards specific locations. In many ways, this approach can be seen as a "cost-benefit" analysis, considering the advantages and disadvantages associated with migration.

Sjaasted (1962) introduced a human capital approach to migration, treating it as an individual investment decision aimed at enhancing human capital productivity. The central idea of this model is that individuals engage in a rational cost-benefit analysis, considering the expected discounted returns of migration over future time periods. Migration occurs only if the expected returns are positive. The model emphasizes the significance of migration costs in the decision-making process. The study identifies two primary migration costs: economic costs and psychological costs. Economic costs refer to the financial expenses associated with migration, while psychological costs encompass the emotional and social adjustments required when moving to a new location. These migration costs are shared between household members, with support provided by the family to the migrant in times of need, such as securing urban employment. In return, the migrant sends remittances back home to assist the family with consumption smoothing and investment activities, such as agricultural investments. As a result, there exists a form of interdependence between the migrant and the family members left behind, with mutual support and contributions.

2.2.2 Family Migration Decision

The scholarly literature on migration acknowledges that families typically play a crucial role in making migration decisions. Instead of focusing on individual desires, the decision-making process within families is often centred around maximizing the overall well-being of the family unit. There are two primary frameworks for family migration decision-making. In the first framework, the entire family chooses to migrate together, while the second framework involves the family acting as a unified decision-making unit that determines whether a member should migrate based on what benefits the family as a whole.

The pioneering work on the role of families in migration decisions was conducted by Sandell (1977) and Mincer (1978), who developed the first comprehensive model in this area. Sandell (1977) specifically explores the concept of women's migration as part of family migration. The study begins with the hypothesis that if only one partner in a couple finds a better job opportunity in a different location, the entire family would migrate. In other words, the decision to migrate is made based on whether the gains of one family member compensate for the losses experienced by the other family member. By applying a legit model to analyse secondary data from the National Longitudinal Surveys (NLS), Sandell's study establishes a positive relationship between education and the likelihood of migration. One notable finding from the research is that the migration rate decreases when the decision involves the migration of the entire family for the purpose of the female member's job. Therefore, the study reveals that working women are disproportionately affected when a family undergoes migration, and it is primarily the female partner who is often tied to the decision to move. Instances where families migrate specifically to the location where the female member works are relatively rare based on the findings of this study.

Mincer's (1978) seminal work is considered a fundamental contribution to the literature on family migration. It was the first model to explicitly differentiate between individual and family migration decisions within a theoretical framework. A key assumption of the model was that, in the context of migration, decisions made by the family take precedence over individual decisions. The study primarily focused on examining the impact of family ties on migration, as well as the consequences of migration decisions on the employment, income, and overall stability of family members. It was discovered that spouses, in their pursuit of maximizing family well-being, often disregard situations that would maximize their individual utility. Hence, the well-being of the family as a whole was given greater importance than the well-being of individual members. The study also uncovered several noteworthy findings. For instance, the presence of school-age children was identified as a barrier to family migration. Additionally, families with a single earner were more likely to migrate compared to those with two earners. Moreover, families with highly educated husbands exhibited a higher propensity for migration, providing support for the notion that human capital plays a significant role in family migration decisions, similar to Sjaastad's (1962) model for individual migration decisions.

Bartel's (1979) significant research on family migration also holds considerable influence. The study delves into various factors that determine family migration and explores the effects of migration on the migrant's life cycle. One notable finding is the strong connection between job mobility and migration. The study reveals an inverse relationship between the existing wage rate and the likelihood of migration. Moreover, the wage obtained after migration is largely influenced by the migrant's age and the type of migration undertaken.

Agesa and Kim (2001) emphasize the household's central role as the primary decision-making unit in individual migration. The study examines two types of migration streams: split migration and family migration. Split migration refers to situations where the head of the household migrates first, followed by other family members at a later point. On the other hand, family migration entails the entire family moving together. The research highlights several factors that significantly influence the decision-making process, including age, the number of children and dependents in the household, human capital skills, job security, and the origin of the migrant household. It is revealed that larger family sizes tend to facilitate family migration, meaning that as the size of the family increases, the likelihood of migrating as a family unit also increases.

Jong (2000) in his study reveals the significant role of "expectations" in migration behaviour for both male and female family members. These expectations pertain to achieving future goals, whether it involves staying in their native place or moving to a different location. Importantly, these expectations differ based on gender. The study employs the theory of planned behaviour to examine the decision-making process of migration, incorporating the element of expectations.

The findings indicate that certain variables, such as marital status, responsibilities for dependent care, and previous migration experience, have a significant impact on migration decisions. Conversely, variables like human capital (education), household income levels, land ownership, family size, and migrant networks were found to be statistically insignificant in influencing migration decisions. The study concludes that expectations, family norms, and gender roles are key determinants in the decision-making process of migration.

Klein (2011) provides an analysis of the demographic differences between migrant families and families that remain in their original location, using a framework centred around fertility decisions. In testing the hypothesis that the real rural-urban income gap is the most decisive factor in migration, the study reveals that this is not the case. Instead, education emerges as the most significant determinant of family migration. The research demonstrates that individuals are more likely to migrate when their siblings reach primary school age. Furthermore, the study finds that families with younger members and smaller sizes are more inclined to migrate compared to other families. However, it also uncovers a noteworthy finding that the level of schooling attained by individuals is inversely related to the duration of their migration period. In other words, education acts as a factor that reduces the length of the migration period, suggesting that it plays a retarding role in migration duration.

2.2.3 Mental Modeling

Craik (1943) explains that if an organism possesses a miniature representation of the external world and its potential actions within its mind, it gains the ability to experiment with different options, determine the most favorable among them, anticipate future circumstances, apply knowledge from past experiences to handle the present and future, and overall respond more effectively, securely, and skillfully to challenging situations.

Gray et al, (2013) reported that the concept and expected application of a collaborative modeling tool named 'Mental Modeler,' which utilizes fuzzy-logic cognitive mapping (FCM), aims to express stakeholders' mental models and enable the integration of diverse knowledge in environmental decision-making. It allows the formulation of hypotheses for testing, as well as the simulation of scenarios to assess the anticipated outcomes of proposed policies.

Martin et al, (2011) elicited that enhanced policies and decisions can be achieved through more comprehensive and dynamic mental models. By eliciting and comparing such models, important variables and their underlying structures can be systematically clarified. Additionally, superior dynamic mental models can be identified. This research paper examines previous studies that assess and compare mental models. It reveals that existing methods used for model comparison fail to consider crucial aspects of dynamic systems, such as time delays in causal links, feedback structures, and the polarities of feedback loops. Mental models lacking these properties are typically static models. To address these limitations, we improve upon the widely used distance ratio approach (Markóczy and Goldberg, 1995) to incorporate these dynamic characteristics and identify differences among mental models at three levels: element level, individual feedback loop level, and complete model level. Our contribution lies in a novel method for comparing explicitly articulated mental models, rather than eliciting such models. The application of this method demonstrates that previously unavailable information is essential for understanding disparities in managers' mental models of dynamic systems.

Westbrook, L (2006) suggests that the term "mental model" and related concepts within the socio-cognitive framework of sense-making, examines the connections and debates surrounding mental model research in information studies, and presents the findings of an exploratory study of the mental models of an academic information system. People create mental models of the systems and processes they engage with as part of the process of making sense of their experiences. These models feature (3) methods for engaging with the system or process, (4) important components, and (5) interactions between those components. Three significant trends are found in the graduate students' models of information seeking in the small-scale study addressed here.

3

Theory and Concepts

Theoretical orientation involves the scholarly and theoretical process that shapes the precise definition and written content of a complete research project. It involves incorporating conceptual ideas, logical discussions, and the quest for knowledge to develop a theory that aligns with the defined objectives and research design established during the research process.

3.1 Migration

Migration refers to the process of individuals or groups of people moving from one place to another with the intention of establishing a new permanent or semi-permanent residence. It involves the relocation of people across geographical, political, or social boundaries, typically over long distances. Migration can occur within a country (internal migration) or involve crossing international borders (international migration).

People migrate for various reasons, including economic opportunities, seeking better living conditions, escaping persecution or conflict, joining family members, pursuing education or career opportunities, or being forced to leave their homes due to natural disasters or environmental factors. The motivations behind migration can be complex and multifaceted, influenced by a combination of push factors (negative circumstances in the home country) and pull factors (attractive opportunities in the destination country).Migration can take different forms, such as labour migration, where individuals move to another region or country for employment purposes, or refugee migration, where people flee their home countries due to fear of persecution or violence. It can be temporary or permanent, with some migrants returning to their home countries after a period of time, while others settle permanently in their new location and may eventually become citizens of that country.

There are several categories under which migrants may be divided, including:

Emigrant: Emigrants are individuals who leave their home country to settle in a different country or region. They are the ones who initiate the migration process by voluntarily choosing to depart from their country of origin. Emigrants typically seek better opportunities, such as improved living conditions, employment prospects, educational options, or a desire for a new cultural experience in the destination country. They undergo the process of emigration, leaving behind their familiar surroundings to establish a new permanent or semi-permanent residence elsewhere.

Immigrant: Immigrants are individuals who move to a new country or region with the intention of establishing a new permanent or semi-permanent residence. They are the ones who arrive in a foreign country or region, often with the intention of settling and integrating into the society of the host country. Immigrants are typically seeking better opportunities, such as economic prospects, educational opportunities, political stability, or to join family members already residing in the destination country. They undergo the process of immigration, which involves obtaining legal permission, such as visas or residency permits, to live and work in the new country. Immigrants contribute to the cultural, social, and economic diversity of their host countries and play a significant role in shaping the fabric of societies.

Refugee: Refugees are individuals who have been forced to leave their home countries due to a well-founded fear of persecution, conflict, violence, or human rights violations. They are unable or unwilling to return to their home country due to the dangers they would face. Refugees seek safety and protection in another country and are granted a special legal status called refugee status.

Unlike other migrants who may move for economic or personal reasons, refugees are compelled to flee due to threats to their lives or freedom. They may be escaping from situations such as armed conflicts, ethnic or religious persecution, political oppression, or severe human rights abuses. The process of seeking refuge is often urgent and necessitates leaving behind homes, possessions, and sometimes even family members. Refugees are protected under international law, particularly the 1951 United Nations Convention relating to the Status of Refugees. They are entitled to certain rights, including the right to seek asylum, protection from refoulement (forced return to their home country), access to basic necessities, and the possibility of resettlement in a safe third country. The host countries where refugees seek asylum are responsible for providing protection, including temporary shelter, access to basic services, and the opportunity to rebuild their lives. Refugee status is usually granted after a thorough assessment of an individual's claim, which involves determining the validity of their fear of persecution and the need for international protection.

Refugee migration is often a complex and challenging process, both for the individuals seeking refuge and for the countries receiving them. It involves significant social, economic, and political implications, and addressing the needs of refugees requires cooperation and support from the international community.

Internally Displaced Person: Internally Displaced Persons (IDPs) are individuals or groups who have been forced to flee their homes due to conflicts, violence, natural disasters, or other circumstances, but remain within the borders of their own country. Unlike refugees, IDPs have not crossed an international border to seek safety and protection.

The term "internally displaced person" is used to distinguish these individuals from other categories of migrants, such as economic migrants or voluntary migrants. IDPs are protected under international humanitarian law and are entitled to certain rights, including the right to receive assistance and protection from their own government.

3.1.1 Causes of migration

Understanding the reasons behind why certain individuals choose to migrate while others do not can be fascinating. The key factors that serve as motivations for people to relocate can be broadly categorized into five groups, which are elaborated on below:

Economic Factors: Despite the importance of non-economic factors, the majority of studies indicate that economic factors are the primary driving force behind migration. In many developing countries, low agricultural income, agricultural unemployment, and underemployment are key factors that push migrants toward more prosperous or dynamic areas offering better job opportunities. Additionally, the pressure of population resulting in a high man-land ratio is widely recognized as a significant cause of poverty and rural outmigration. Thus, most studies agree that the main reason for migration (excluding forced and sequential migration) is the pursuit of improved economic prospects. This holds true for both internal and international migration. The economic factors that motivate migration can be further divided into "Push Factors" and "Pull Factors." In other words, people migrate either due to compelling circumstances that force them to leave their place of origin or because they are attracted by favourable conditions in a new location.

Push factors are circumstances or conditions that compel individuals to leave their current location and seek opportunities elsewhere. These factors include low productivity, unemployment, underdevelopment, poor economic conditions, lack of advancement opportunities, depletion of natural re*Source*s, and natural disasters. In many developing countries, rapid population growth has led to a decline in the land-to-population ratio, resulting in increased unemployment and underemployment. The introduction of capital-intensive agricultural methods and mechanization also reduce the need for labour in rural areas. The lack of alternative *Source*s of income in rural areas, such as non-agricultural activities, is another significant factor driving migration. Additionally, the persistence of joint family systems and inheritance laws that prohibit property division may force many young men to migrate to cities in search of employment. Even the sub-division of property can lead to migration when the property becomes too small to support a family.

On the other hand, pull factors are factors that attract migrants to a particular area. These factors include better employment opportunities, higher wages, facilities, improved working conditions, and amenities. City ward migration is often observed when there is rapid industrial, commercial, and business growth. The process of industrialization, technological advancements, and cultural changes in modern society worldwide are closely related to migration from rural to urban areas. Under the capitalist model of development, a significant proportion of investments concentrate in urban centres, encouraging people to move there in search of higher-paying jobs. In recent years, the high migration rates from India and other developing countries to the UK, USA, Canada, and the Middle East can be attributed to better employment opportunities, higher wages, and the potential for a higher standard of living. Sometimes, people are also attracted to cities due to better cultural and entertainment activities. Pull factors are influential not only in rural-urban migration but also in other forms of domestic and international migration.

The question of whether push or pull factors are more important is debated among researchers. Some argue that push factors carry more weight, as they believe rural problems have a dominant role in driving population migration. Conversely, proponents of pull factors consider them more significant, emphasizing the high rates of investment in urban areas, which create more employment and business opportunities and a greater allure for urban lifestyles. There is also a third group of researchers who argue that push and pull factors are closely interconnected. Those who are pushed to migrate are simultaneously pulled by the expectation of finding better conditions elsewhere.

Demographic Factors: The variation in population growth rates among different regions within a country has been identified as a determining factor in internal migration. Rural areas tend to have higher fertility rates and natural population increases, which lead to a migration flow from rural to urban areas. In countries like Canada, the United States, New Zealand, Australia, and Gulf countries, immigration has been driven by the shortage of domestic labour supply. Another significant demographic factor influencing internal migration is marriage. Female migration is predominantly associated with marriage, as it is customary in Hindu culture to bring brides from other villages. According to the National Sample Survey, over 46 per cent of migration to urban areas is attributed to marriage. Additionally, the tradition of women returning to their parental homes to give birth to their first child also contributes to significant internal migration.

Socio-Cultural Factors: Social and cultural factors hold significance in the process of migration as well. In some cases, family conflicts and the desire for independence can be driving forces behind migration, particularly among the younger generation. The availability of enhanced communication facilities, such as improved transportation systems and the influence of media *Source*s like television and cinema, along with urban-focused education, can also contribute to migration by influencing attitudes, values, and aspirations.

Political Factors: Political factors can also play a role in either encouraging or discouraging migration between regions. For example, in India, state governments implementing policies like job reservations for local residents, known as the "sons of the soil policy," can deter migration from other states. Nair (1978) noted that 14.5 per cent of individuals who left their native places in Tamil Nadu due to a lack of job opportunities attributed this issue to the anti-Brahmanism propagated by the D.M.K. government in Tamil Nadu. Therefore, the political context, attitudes, and individual perspectives of people have a significant influence on migration patterns.

Miscellaneous Factors: Furthermore, several other factors contribute to migration, including the presence of relatives and friends in urban areas who can offer support and assistance. The desire to access education, which may only be available in urban areas, is also a motivating factor for migration. Cultural proximity, cultural diversity, and the strength of cultural connections can significantly influence migration patterns. Personality traits such as high vitality, strong self-assertion, and an individualistic attitude are often associated with a higher inclination to migrate.

3.2 Migration Theories

Migration theories provide frameworks for understanding the reasons, patterns, and consequences of migration. Here are some notable migration theories:

3.2.1 Neo-Classical and Developmentalist Theory

The neoclassical theory of migration views migration as a way to efficiently allocate production factors for the benefit of both the countries of origin and destination. It suggests that the reallocation of labour from rural, agricultural areas to urban, industrial sectors, whether within a country or across borders, is essential for economic growth and overall development (Todaro, 1969). According to this theory, as labour freely moves in an unrestricted market environment, it results in a scarcity of labour and consequently increases the productivity of labour and wages in the countries that send migrants. Conversely, capital flows are expected to move in the opposite direction, from countries with a scarcity of labour to those with a scarcity of capital. Ultimately, this process of equalizing factor prices, as described by the Heckscher-Ohlin model, predicts that migration will cease once wage levels become similar in both the origin and destination countries (Massey*et al.*, 1998).

In a pure neoclassical framework, the developmental impact of migration is solely achieved through the equalization of factor prices. As Djajic (1986) pointed out, traditional neoclassical migration theory did not consider the possibility of benefits for non-migrants. Specifically, neoclassical migration theory does not account for the flow of money remittances back to the countries of origin (Taylor, 1999:65). Neoclassical migration theory typically perceives migrants as independent individuals driven by maximizing their own utility and tends to overlook other motivations for migration, as well as the importance of social groups such as households, families, and communities to migrants.

During the 1950s and 1960s, prevalent perspectives in development theory regarded returning migrants as significant catalysts for change and progress. It was anticipated that migrants would not only bring financial re*Source*s but also introduce fresh ideas, knowledge, and entrepreneurial mind-sets. As a result, migrants were seen as having a positive impact on development by facilitating the rapid spread of modernization in developing nations. Additionally, remittances were recognized as playing a crucial role in stimulating economic growth. Remittances were widely recognized as a crucial *Source* of foreign currency at the macro level. At the meso and micro levels, migration was expected to bring about economic improvements in the regions from where migrants originated. Remittances were believed to enhance income distribution and improve the overall quality of life in ways that other development approaches could not achieve (Keely and Tran, 1989:500). Additionally, it was anticipated that labour migrants or "guest-workers" would make significant reinvestments in businesses and enterprises in their home countries upon their anticipated return. Migrant workers were seen as a beacon of hope for the industrial development of their countries of origin (Beijer, 1970:102), and it was widely believed that large-scale emigration could contribute to the best outcomes for both the immigration country in terms of rapid growth and the origin country in terms of its own rapid growth (Kindleberger, 1965:253).

3.2.2 Cumulative Causation Theory and the "Migrant Syndrome"

Myrdal (1957) elaborated on the theory of cumulative causation, which suggests that capitalist development inevitably leads to widening spatial disparities in well-being. Once differential growth takes place, internal and external economies of scale, such as agglomeration and multiplier effects, perpetuate and deepen the polarized pattern characterized by a cycle of poverty in peripheral regions and accelerated growth in core regions. As a result, economic activities in areas and countries with an initial advantage attract investment and encourage the emigration of highly skilled populations from peripheral areas and countries. While there may be positive "spread effects," such as increased demand for agricultural products or raw materials trade from the periphery, these positive effects are outweighed by the negative "backwash effects." Myrdal argued that, without robust state policies, the capitalist system exacerbates spatial inequalities.

Cumulative causation theory is applicable at both national and international levels, and it shares similarities with centre-periphery models and neo-Marxist development theory. Therefore, it aligns well with historical-structuralist perspectives on migration and development. According to this theory, migration is anticipated to have negative effects on regional and national economies as it depletes them of valuable human and material capital re*Source*s. These re*Source*s are exploited for the benefit of industrialized countries in the case of international migration, and for urban-based capitalist elite groups within developing countries in the case of internal migration, who seek inexpensive migrant labour. Migration undermines regional and local economies by stripping communities of their most valuable labour force, leading to increased reliance on core countries (with remittances being just one aspect) and stimulating further out-migration. Consequently, the productive structures in the countries of origin gradually erode, resulting in "asymmetric growth" instead of conforming to the neoclassical equilibrium model of factor price equalization. This process deepens the underdevelopment and dependency of underdeveloped regions on developed core countries (Almeida, 1973). Moreover, this pauperization is seen as a driver for additional out-migration. Implicit in this perspective is the assumption that migration is more or less directly influenced by spatial disparities in opportunities, underdevelopment, and poverty.

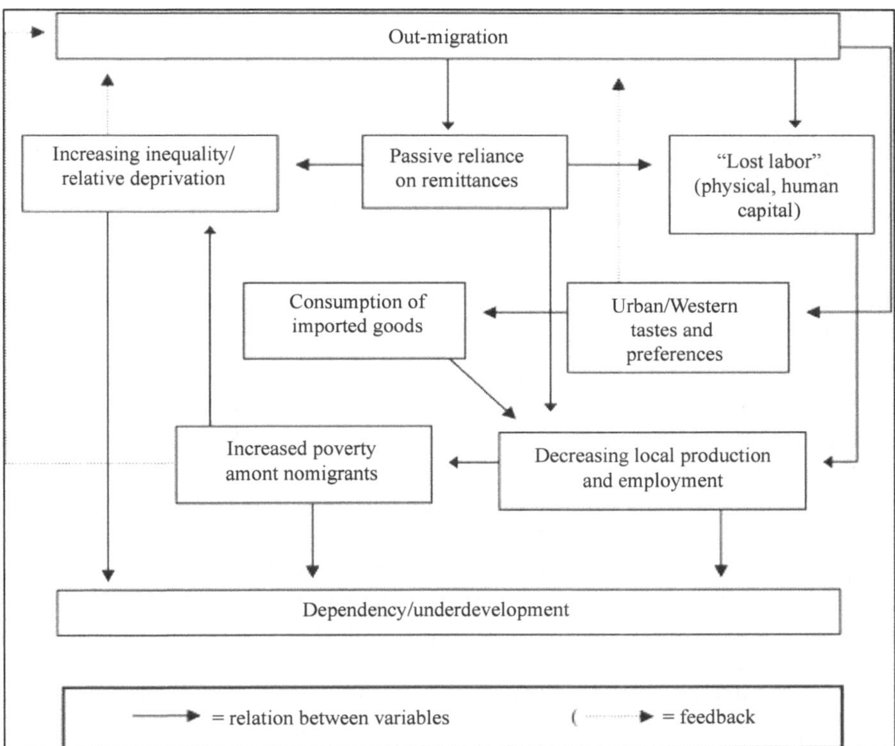

Fig. 3.2.2.1: Conceptual Framework of the "Migrant Syndrome"

Source: De Haas H.(2010), Migration and Development: A Theoretical Perspective.

3.2.3 New Economics of Labour Migration (NELM)

During the 1980s and 1990s, the new economics of labour migration (NELM) emerged as a significant critique of the neoclassical theory of migration (Massey *et al.*, 1993:436). The NELM theory challenges the neoclassical models, which often overlook constraints and are deemed too inflexible to address the diverse realities of migration and its interactions with development. One influential figure in revitalizing the understanding of migration in and from the developing world was Stark (1978, 1991), who emphasized placing individual migrants' behaviour within a broader societal context. Stark also advocated for considering the family or household as the most appropriate decision-making unit, rather than focusing solely on the individual. This new approach broadens the scope to incorporate factors beyond individual utility maximization that influence migration decision-making.

The new economics of labour migration conceptualizes migration as a form of risk-sharing behaviour among families or households. In comparison to individuals, households are considered better equipped to diversify their re*Source*s, including labour, in order to minimize the risks associated with income fluctuations (Stark and Levhari, 1982). The underlying assumption is that people, households, and families not only aim to maximize income but also seek to minimize and distribute risks. Both internal and international migration can thus be seen as a response by households to income risks, with migrant remittances serving as a form of income insurance for the households left behind. This risk-sharing motive can even explain instances of migration in the absence of expected wage differentials. The notion is that, from the perspective of the entire household, it may be a strategy that improves the well-being of all members to have some members migrate elsewhere, whether as a means of sharing risks or as an investment in accessing higher earning opportunities (Lucas and Stark, 1985:902).

Migration is not only viewed as a risk-spreading strategy for households but also as a means to overcome various constraints present in imperfect credit and risk markets, which are prevalent in many developing countries (Stark and Levhari, 1982; Stark and Bloom, 1985; Taylor, 1986; Taylor and Wyatt, 1996; Taylor, 1999). These markets are often underdeveloped or inaccessible for non-elite groups. Migration, particularly through international remittances, can serve as a household strategy to bypass these market limitations by allowing households to invest in productive activities and improve their overall well-being (Stark, 1980). Unlike in neoclassical migration theory, which tends to overlook remittances, the new economics of labour migration acknowledges them as one of the primary motives for migration. Conceptually, this also implies that the developmental impact of migrants is not solely dependent on return migration. Migrants who remain abroad, become permanent settlers, or their descendants can contribute to development through the remittance of money.

3.2.4 Dual Labour Market Theory

In 1979, Michael J. Piore introduced the Dual Labour Market Theory, which deviates from individual-focused models. Instead of viewing migration as a result of individual decisions, this theory argues that international migration is primarily driven by the intrinsic labour demands of industrialized societies. Piore (1979) highlights that the permanent demand for labour in industrialized and developed nations is a key factor in international migration, serving their development agendas.

Essentially, international migration occurs not primarily due to push factors in sending countries, but because of pull factors in receiving countries. Push factors include low wages and high unemployment, while pull factors involve the essential and unavoidable needs that foreign workers are expected to fulfill in receiving countries. The theory also emphasizes four main characteristics of industrialized countries that explain why they attract labour from other nations: structural inflation, motivational problems, economic dualism, and the demographics of the labour supply (Massey *et al.*, 1993).

The Dual Labour Market Theory presents implications that differ from macro-level models, although it remains consistent with the fundamental concepts of Neo-classical Economics (Massey *et al.*, 1993). One significant implication, which sets it apart from Neo-classical and NELM (New Economics of Labour Migration) approaches, is its assertion that international labour migration is driven by demand. According to this theory, the demand for migrant workers arises from the structural requirements of the economy, rather than being determined solely by wage disparities or the preferences of households or families.

3.2.5 Social Network Theory

Labour migration can occur for a variety of reasons. These reasons include the pursuit of higher individual income, the attempt to diversify household income risks, international displacement with a strategy for market penetration, and recruitment programs aimed at meeting employers' demand for low-wage workers (Massey *et al.*, 1993). However, while these reasons can be observed, they alone are insufficient to explain actual migration patterns. Other factors such as geographical proximity to nation-states, the availability of social networks, institutions, and cultural and historical factors need to be considered (De Haas, 2010b). Migration networks, which are linked to the concept of social capital, play a crucial role. Arango (2000) defines migration networks as interpersonal connections that link migrants with relatives, friends, or fellow countrymen who provide information, financial support, employment opportunities, and assistance with accommodation in various ways. These networks reduce the costs and risks of migration and increase the expected benefits (Massey *et al.,* 1993). As a result of these networks, subsequent migrations have positively contributed to enhancing opportunities for other migrants in their decision-making process. Furthermore, Vertovec (2002) and Dustmann and Glitz (2005) assert that the diaspora and other networks have the ability to influence migrants when they choose their destinations. It is evident that network connections serve as a form of social capital, providing broad access to employment opportunities abroad (Massey *et al.*, 1993).

As a result, the positive effects of the migration network mechanism have had an impact on the development of sending countries, while also giving rise to significant legal, political, and financial challenges regarding immigration in the receiving countries. Van Naerssen, *et al.* (2008) have identified "transnational communities" as mechanisms that perpetuate their own sociocultural practices in foreign lands, creating an extended national market and influencing the development of migrant businesses in both the origin and host countries. Consequently, these multinational or transnational businesses have facilitated the exchange of various commodities, capital, ideas, and skills, thereby contributing to the development of social, political, cultural, and economic ties (Prakash, 2009).

3.2.6 Household Livelihood Strategy on Migration

The new economics of labour migration exhibits notable conceptual similarities, although not yet fully explored, with livelihood approaches. Livelihood approaches emerged in the late 1970s among geographers, anthropologists, and sociologists conducting micro-level research in developing countries. They observed that the varied and sometimes conflicting findings from their empirical work did not neatly align with rigid neo-Marxist frameworks. Consequently, they argued that the impoverished population should not be solely portrayed as passive victims of global capitalist forces but as individuals who actively strive to enhance their livelihoods within the constraints of their circumstances. This highlights the crucial role of human agency in shaping livelihood outcomes (Lieten and Nieuwenhuys, 1989).

A livelihood refers to the combination of capabilities, assets (including both material and social re*Sources*), and activities necessary for sustaining a means of living (Carney, 1998). It encompasses not only the income-generating activities of households but also the social institutions, intra-household relationships, and mechanisms through which individuals access re*Source*s throughout their lives (Ellis, 1998). A livelihood strategy can be defined as a deliberate and strategic choice made by households and their members to maintain, secure, and improve their livelihoods. This choice is influenced by factors such as selective access to assets, perceptions of opportunities, and the aspirations of individuals. As these factors vary across households and individuals, livelihood strategies exhibit considerable heterogeneity.

The introduction of the livelihood concept marks a shift away from rigid and deductive historical-structuralist perspectives towards more empirical approaches. This shift recognizes that people, particularly in circumstances of economic, political, and environmental uncertainty and hardship, organize their livelihoods within broader social contexts, such as households, village communities, and ethnic groups. The household, in particular, has been acknowledged as the most appropriate unit of analysis for many social contexts (McDowell and De Haan, 1997:3).

In this context, the role of migration in diversifying, securing, and potentially enhancing household livelihoods has gained increasing recognition. Migration is often part of a broader set of strategies employed by households, which may include agricultural intensification and engagement in local non-farm activities (McDowell and De Haan, 1997; Bebbington, 1999; Ellis, 2000). It is now acknowledged that migration goes beyond being a short-term survival tactic for rural populations who have been uprooted by global capitalist forces and pushed into the ranks of an emerging international proletariat. Empirical studies suggest that migration is often a purposeful decision aimed at improving livelihoods, facilitating investments (Bebbington, 1999:2027), and mitigating income fluctuations that were previously heavily influenced by climatic uncertainties (McDowell and De Haan, 1997:18; De Haan*et al.*, 2000:28). Migration can be viewed as a means to acquire a broader range of assets that act as insurance against future shocks and stresses (De Haan*et al.*, 2000:30). Although this argument has primarily been applied to rural-urban internal migration in impoverished countries, there is no inherent reason why the concept of diversification through migration cannot also be extended to international migration and urban households.

Therefore, from a household perspective, it is not a matter of choosing between migration or activities at the origin, but rather a combination of both. This highlights the need to consider the interplay between migration and other diverse livelihood strategies across different sectors and locations, encompassing the entirety of the household's activity "portfolio" (Stark, 1991). Research that an attempt to isolate migration and migrants from their broader social and economic context is unable to fully comprehend the relationship between migration and the broader transformative processes encapsulated in the concept of development.

3.2.7 Transnationalism

The emergence of new economics and livelihood perspectives in the study of migration and development has coincided with a third trend known as the "transnational turn" in the examination of migrant communities' settlement and integration in receiving countries (Glick Schiller, Basch, and Blanc-Szanton, 1991; Castles and Miller, 2009; Faist, 2004). There is a growing recognition of the increased opportunities for migrants and their families to live transnationally and embrace transnational identities (Vertovec, 1999; Guarnizo, Portes, and Haller, 2003). This is facilitated by advancements in communication technology, such as mobile phones, fax, satellite television, and the internet, as well as globalized formal and informal banking systems that enable the transfer of remittances. These developments allow migrants and their families to maintain double loyalties, travel back and forth, connect with people, work, and conduct business simultaneously across different locations. While it is true that migrants in the 19th and early 20th centuries also maintained transnational ties, it is likely that technological advancements have significantly expanded the opportunities for migrants and their families to pursue transnational livelihoods on a more constant, day-to-day basis.

The increasing transnational nature of migrants' lives has posed challenges to assimilationist models of migrant integration and the traditional understanding of nation-states and citizenship. As a result, clear distinctions between "origin" and "destination" and categories like "permanent," "temporary," and "return" migration are becoming harder to maintain in a world where migrants engage in circulation and maintain simultaneous commitments to multiple societies or communities (De Haas, 2005).

This has significant implications for the study of migration and development, as it challenges the notion that integration into receiving societies and commitment to origin societies are mutually exclusive. Previously, it was commonly assumed that migrants' integration would lead to a gradual weakening of ties with their societies of origin, and that "permanent" migration would therefore result in a loss or drain for the origin countries. This assumption largely contributed to skepticism regarding the sustainability of remittances and the belief that migrants' contributions to development in their home countries relied heavily on return migration. However, empirical research has shown that migrants can maintain strong transnational ties over extended periods of time, and these ties can even extend across multiple generations. Studies have also demonstrated that migrants' engagement with their origin countries is not contingent upon their return; rather, it can be sustained through remittances, communication technologies, visits during holidays, and patterns of pendular migration.

The enduring nature of transnational ties can be observed through continuous remittances, transnational marriages, and the active participation of migrants in the social, cultural, and political affairs of their home countries. Therefore, it would be incorrect to automatically interpret migrants' commitment to their countries of origin as a sign of failed integration. Conversely, migrants' deep involvement in their receiving societies does not necessarily diminish their significant commitment to their home countries (Snel, Engbersen, and Leerkes, 2006). In fact, it is possible that successfully "integrated" migrants possess increased financial and human re*Source*s, which can potentially enable them to establish businesses or engage in public discourse in their origin countries.

3.2.8 Social Ecology Theory

Social ecology theory, also known as ecological systems theory, is a theoretical framework developed by psychologist UrieBronfenbrenner. It focuses on the interactions between individuals and their social, cultural, and physical environments, emphasizing the reciprocal influence between people and their ecological contexts. Social ecology theory recognizes that human development is shaped by multiple systems that range from immediate, personal contexts to broader societal and cultural influences. The key components of social ecology theory are:

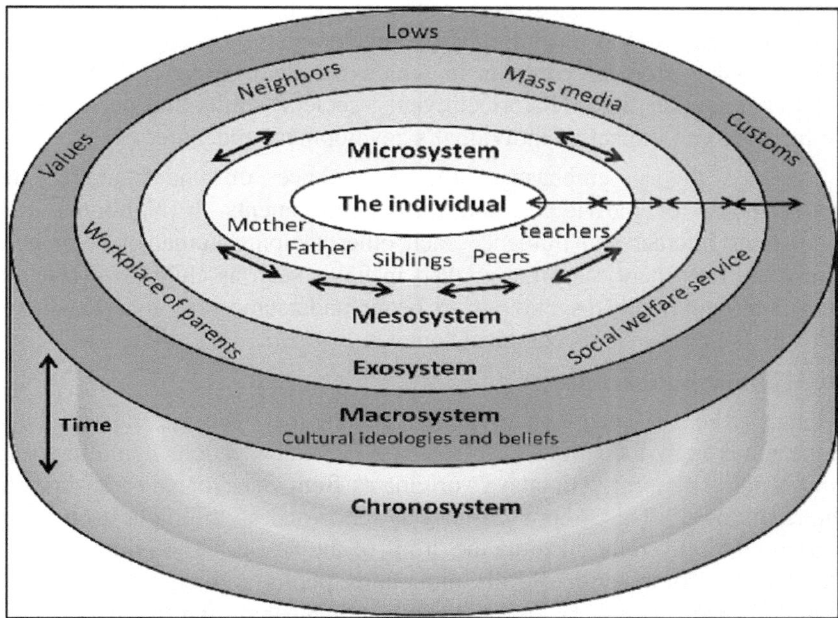

Fig. 3.2.8.1: Social Ecology components

1. **Microsystem:** The microsystem refers to the immediate environments in which an individual directly interacts, such as family, school, and peer groups. These close relationships have a significant impact on a person's development and can influence their beliefs, attitudes, and behaviours.
2. **Mesosystem**: The mesosystem consists of the connections and interactions between the various components of the microsystem. For example, the relationship between a child's family and school or between their parents and their friends' parents. These interactions can shape the individual's experiences and opportunities for development.
3. **Exosystem:** The exosystem includes settings that indirectly influence an individual's development, even though they may not directly participate in them. Examples of the exosystem include the parent's workplace, community re*Source*s, or local government policies. These external factors can still have significant impacts on an individual's development and well-being.
4. **Macrosystem:** The macrosystem encompasses the broader cultural, social, and economic contexts in which an individual lives. This includes cultural norms, values, laws, and socioeconomic conditions. The macrosystem influences the other systems within social ecology theory and can shape individuals' opportunities and constraints.

5. **Chronosystem:** The chronosystem accounts for the influence of time and historical context on an individual's development. It recognizes that the social and ecological contexts in which people live are not static and can change over time. Historical events, societal shifts, and personal life transitions can impact an individual's development and experiences.

Social ecology theory emphasizes the importance of understanding the interconnectedness of individuals and their environments. It highlights how various systems interact and influence each other, shaping human development and behaviour. This theory is often applied in fields such as child development, education, and community psychology to better understand how individuals are influenced by their social and ecological contexts.

3.3 The Himalayan terrains

The Himalayas consist of a vast mountain system that includes the Himalaya proper, the Karakoram, the Hindu Kush, and several smaller ranges stemming from the Pamir Knot. The term "Himalaya" originates from Sanskrit and translates to "the abode of snow." This mountain range is renowned as the highest on Earth and is often culturally referred to as the "roof of the world." The Himalayas are characterized by a combination of hills, mountains, plains, and large watercourses. Their immense size, ruggedness, and challenging climate make it difficult to explore them in great detail.

Stretching approximately 2,900 kilometres across India, Pakistan, China, and Nepal, the Himalayas stand as the tallest mountain range in the world. Alongside Mount Everest, which remains the tallest peak at 8,848 meters, the range also features several other peaks surpassing 8,000 meters in height. It is the only mountain range known to harbour mountains exceeding 8,000 meters. The origin of the Himalayas can be understood through the "Theory of Continental Drift" proposed by German meteorologist Alfred Wegener. This theory explores the concept of Pangaea, tectonic plates, and the movement of continents toward or away from one another. According to this theory, the Himalayas emerged when Pangaea began to break apart around 200 million years ago, causing the Indian subcontinent to move northward. Approximately 80 million years ago, India was located around 6,400 kilometres south of the Asian landmass and was moving northward at a rate of about 9 meters per century. When India collided with Asia around 40 to 50 million years ago, its northward movement slowed by about half.

The collision between the plates, along with the decrease in their movement speed, is believed to mark the beginning of the rapid uplift of the Himalayas. Geologically, the Himalayas are considered a complex mountain system, with varying width ranging from 400 to 150 kilometres. The entire mountain range is divided into three primary regions:

1. The Greater Himalayas, also known as the Himadri, form the northernmost range and consist of the most continuous series of peaks, with an average

height of around 6,000 meters above sea level. This section is often referred to as the "roof of the world." The folds in the Great Himalayas are irregular in nature, and the core of this part is composed of rock. Due to the high elevations, the peaks in this range are permanently covered with snow and ice. Notable glaciers like the Siachen Glacier, Gangotri, and Yamunotri are found here.

2. The Lesser Himalayas lie south of the Great Himalayas and have average peaks reaching heights of up to 4,000 meters. The heights of the peaks in this range range from 3,700 to 4,500 meters. This region primarily consists of highly compressed and altered rocks. The PirPanjal mountain range is the most significant range in this area and is also the longest. It is home to notable valleys such as the Kashmir Valley, Kangra Valley, and Kullu Valley. This region is well-known for its hill stations, including popular destinations such as Kullu-Manali, Kufri, Shimla, Mussoorie, Nainital, and Darjeeling.
3. The outermost range of the Himalayas is called the Shiwaliks, also known as Manak Parbat in ancient times. Currently, the elevation of this range fluctuates between 900 and 1100 meters. These areas consist of unconsolidated sediments brought by rivers. The longitudinal valleys situated between the Himachal region and the Shiwaliks are referred to as "Duns." Unlike typical river valleys, duns have a secondary *Source* and are protected by rocks and boulders derived from the erosion of the Himalayas and the Shiwalik uplands. Some significant duns in the region include Dehra, Kothri, Chaukham, Patli, and Kota.
4. The Eastern Hills, also known as the Purvanchal, cover the northernmost parts of the North Eastern states and North Bengal. They are primarily composed of sandstones, which are sedimentary rocks. These slopes are densely covered with thick forests. The Patkai Hills, Naga Hills, Manipuri Hills, and Mizo Hills are part of the Eastern Hills region.

3.3.1 Terrains of Darjeeling Himalaya

Darjeeling is situated in the lower region of the Himalayas, characterized by long and rugged mountain ranges that generally extend from north to south. The Sub-Himalayan area consists of separate ridges found in the middle of the Himalayas, while the northern boundaries of the district are marked by snowy mountain ranges that stretch from east to west, approximately 50 miles apart on average. These mountains form the prominent core of the Darjeeling Himalayas. The slopes of the Darjeeling District lie between the smaller Himalayan ranges and the Sub-Himalayan belts of the Eastern Himalayas. The district of Darjeeling is located between 26°31' and 27°13' north latitude and 87°59' and 88°53' east longitude. It is surrounded by the Sikkim Himalayas to the north, the Bhutan Himalayas to the east, and the Nepal Himalayas to the west. The southern foothill area is defined by extensively eroded terraces that stretch in an east-west direction.

The Darjeeling Himalaya is positioned between the Bhutan Himalaya in the east and the Central (Nepal) Himalaya in the west. Geological studies in Sikkim and the neighbouring Bengal region commenced during the mid-19th century. Dr. Hooker, in his renowned work "Himalayan Journals" (1854), documented the geological findings from his extensive surveys in the area. He identified the regional presence of gneisses and observed the overlying layers of sedimentary rocks. The first comprehensive and systematic geological examination of this region was conducted by F.R. Mallet in 1874. Since then, numerous researchers have made additional observations (Dutta, 1950; Gansser, 1964; Acharya, 1972; Powde and Saha, 1982; etc.).

The foothills of the sub-Himalayan regions within the Darjeeling district consist of an area known as Dooars. Dooars is located between 26°16' to 27°.0' N latitude and 88°.4' to 89°.53' E longitude and has an irregular rectangular shape. It has an average elevation of 180 meters above sea level. The area begins from the eastern bank of the Teesta River in the Jalpaiguri district and extends to the western bank of the Sankosh River in the Alipurduar district. Dooars, a region characterized by endless beauty and abundant tea gardens, is situated between the Jalpaiguri and Alipurduar districts in the northern part of the state of West Bengal.

3.3.2 Physiographical Demarcations

The environmental perception of the Darjeeling hill areas sets them apart as unique. The physiographical features of the Darjeeling Himalaya exhibit a high level of complexity, with a wide range of micro and macro relief forms. The hills rise abruptly from the plains, starting at around 150 meters, and the elevation increases as you move northwest towards Sandakphu, reaching a height of 3636 meters. The geomorphic structure of this hilly region is influenced by both geological factors and fluvial processes. Additionally, slope-wash, including mass movements and related phenomena, significantly contribute to the shaping of the landforms. The area is characterized by an abundance of ridges and valleys, forming a mosaic of micro-topographical units. These can be broadly categorized as follows:

i. **Lower hills:** This landscape is characterized by narrow ridges, deep incisions, and numerous scars from mass movements. The average slope in this zone ranges from 10° to 30°, with some slopes extending over 800 meters. These lower hills experience the highest levels of dissection and erosion within the Darjeeling Himalaya.

ii. **Middle hills:** This zone is relatively narrow, situated between the upper and lower hills, and delimited by the 800-meter and 1400-meter contours. The slopes of these hills are primarily occupied by tea plantations.

iii. **Upper hills:** The upper hills are located above the 1400-meter contour and experience extreme climate conditions.

iv. **Foothills:** The foothills of the sub-Himalayan regions in the Darjeeling district encompass Dooars, which span from 90 meters to 1750 meters in elevation.

The variations in the Himalayan terrain result in diverse changes in its physical, physiological, ecological, and sociological aspects as altitude shifts. There are noticeable distinctions in the lifestyle, occupations, and living standards of the people that are influenced by the physical characteristics when observing changes in topographical altitude. By looking beyond natural barriers and environmental changes, we can gain sociological and economic insights into the social environment of the inhabitants.

3.3.3 Historical Background

Darjeeling is believed to have derived its name from 'Dorjeling,' which signifies the place of the *Dorje*, the magnificent thunderbolt of the Lamaistic Religion. Prior to 1706, Darjeeling, along with a portion of Kalimpong and Dooars (Terai), was under the rule of the King of Sikkim. However, in that same year, the Bhutanese seized the present-day Kalimpong area. In the mid-18th century, the Gorkhas conquered Sikkim and the Terai. While the East India Company was focused on expanding their territories in the south, Nepal had already established itself as a powerful nation, extending its areas from Teesta to Kangra, along with various plain regions, by 1800.The Sugauli Treaty was signed in 1816, resulting in Nepal ceding Sikkim and its Terai region to the British. Shortly after, the Treaty of Titleya followed, and on February 10, 1817, Sikkim and the Terai were returned to the King of Sikkim, making it a buffer state. At this time, Darjeeling (proper) was a small village under the Kazi (minister) of the King of Sikkim. In 1827, Captain Lloyd and Grant visited Darjeeling while settling a dispute between Nepal and Sikkim, and they discovered its revitalizing climate. The favourable weather conditions led to the establishment of the winter capital and military base.

With Lloyd's efforts, the Governor General Lord Bentick initiated negotiations with the King of Sikkim. On February 1, 1835, the King of Sikkim presented Darjeeling to the British Government as a token of friendship, considering it a worthless uninhabited mountain, with the intention of allowing British officials to benefit from its healthy climate. In 1836, Darjeeling was a small market consisting of a few houses with mat roofs, commonly known as GOONDRI BAZAR. Kalimpong, which was previously part of Bhutan, was incorporated into Darjeeling by the British. Additionally, in 1880, the Siliguri area was separated from Jalpaiguri district and included in the Darjeeling District. The geographical shape of Darjeeling remained unchanged, but its political affiliations varied over time.

After 1850, the district of Darjeeling was annexed to Rajshahi (now in Bangladesh). In 1905, the region, in its present form, became part of Bhagalpur in Bihar (when Bengal, Bihar, and Orissa were undivided). However, it was transferred back to Rajshahi in 1912 and eventually became part of Bengal after gaining independence.

3.3.4 Society

Darjeeling Himalaya represents a region characterized by its diverse ethnicity, culture, and language. The society in this area comprises various elements originating from different backgrounds. The social diversity found here is a significant and powerful aspect of the region. Different social groups with diverse ethnic and linguistic origins, representing various racial backgrounds and social statuses, have established themselves in different ecological niches provided by the area's physical geography and climate. The waves of immigration have brought the ancestors of the majority of the current population from neighbouring regions across the Himalayas. As a result, a social mosaic with distinct ethnic characteristics has been formed.

The population composition of Darjeeling district can be roughly categorized into the following ethnic groups:

- Nepalese: This term encompasses more than 15 ethnic groups. Various castes and tribes, including Sherpas, who migrated to the area from Nepal during the 19th and early 20th centuries, fall under this category.
- Lepcha: These are the indigenous tribes of the area.
- Bhutia: This tribal group includes both Bhutanese and SikkimeseBhutias.
- Tibetan: These are refugees who arrived in the area after 1961.
- Bengali: This category includes both permanent residents and migrants from southern Bengal, as well as refugees from Bangladesh.
- Other Indians: This refers to individuals from various other regions of India residing in the area.

The population of Darjeeling district exhibits a rich tapestry of ethnic diversity, with each group contributing to the cultural fabric of the region.

Over time, this scenario has resulted in political and socioeconomic tensions between the indigenous population and those from outside the region. The Himalayan area experiences occasional severe ethnic conflicts, and there is a persistent demand for an independent region called Gorkhaland. The majority of people in the area, who are mostly Nepalese, desire a distinct identity separate from West Bengal.

3.3.5 Economy

The livelihood and economy of the people in the Darjeeling Himalaya are heavily influenced by environmental conditions. The region's topography, climate variations, and soil conditions have all played a role in shaping human settlements. Rural communities primarily engage in subsistence agriculture, livestock rearing, forestry, plantations, and related activities. However, it is worth noting that the Darjeeling Himalaya has a relatively small percentage (only 13%) of cultivable land compared to its total geographical area. Unfortunately, the government has not made any efforts to maximize the utilization of this land.

The physiography of the area results in highly diverse agricultural conditions. Altitude and slope aspect greatly impact agricultural practices. Due to the cold climate, crops are not cultivated above an altitude of 9500 feet. This challenging situation in the Darjeeling Himalaya makes agriculture extremely difficult and requires significant human labour input. A large portion of people residing in Khasmal areas still practice traditional agriculture, accounting for more than 35% of the land falling under this category. The presence of bare rocky terrain and steep slopes greatly restrict agricultural operations in the region.

In the Himalaya, agricultural crops can be broadly categorized into two groups: food crops and cash crops. Food crops consist of rice, maize, potato, wheat, barley, and others, while cash crops include tea, cinchona, ginger, and similar crops. The agricultural methods employed vary depending on the type of crop being cultivated. Valleys and low-altitude areas are suitable for growing fruits such as oranges, papayas, peaches, guavas, plums, and even mangoes. Livestock rearing and animal husbandry play a significant role in the rural communities of the region. Grazing of livestock is a common practice in the valleys of the Darjeeling Himalaya.

Agriculture is typically practiced on irrigated terraces known as "Khets" or rain-fed terraces called "Baris." Khets are primarily found at lower altitudes (below approximately 1500 meters) and are mainly used for paddy cultivation, although winter or dry season crops are also grown in many cases. On the other hand, Baris are prevalent at both lower and higher altitudes (up to approximately 3500 meters). As the altitude increases, the proportion of Baris to Khets also increases due to cooler dry season conditions, steeper slopes, and limited water accessibility.

Livestock provides draft power and serves as the primary and most important *Source* of fertilizer. Villagers use pairs of oxen to cultivate the terraced fields. Animal manure is essential for field fertilization as artificial fertilizers are rarely used.

The Darjeeling Himalaya region is renowned for its tea, tourism, and timber industries. It is also a significant base for cultivating cinchona and other medicinal plants. Over time, sericulture has been expanding in the area as well. The Darjeeling Himalaya is globally recognized for producing the finest quality tea, which commands the highest prices. Tea cultivation began on a commercial scale in 1856, and presently, there are tea gardens spanning 19,239 hectares, yielding 11-12 million kilograms of tea annually. Cinchona was introduced for trial cultivation between 1861 and 1869, with large-scale production commencing in 1887. The Darjeeling Hills are known for producing top-quality Indian mulberry silk. Forestry is a significant occupation in the region, with approximately 38.91% of the area covered by forests. Numerous forest-based industries have emerged, indicating substantial potential for further development. The cultivation of aromatic medicines, exotic plants, and orchids also contributes to the local economy. Tourism is a highly promising industry in Darjeeling. It has been a popular destination for tourists throughout history and continues to attract a large number of visitors from all over the country and even abroad.

Over time, the significant human and livestock population in the Himalayas, much like in other parts of the country, has resulted in extensive destruction of both renewable and non-renewable re*Source*s in this delicate region. Additionally, detrimental agricultural practices, overgrazing of hill slopes in catchment areas, deforestation for fuel, fodder, timber, and construction materials, all contribute to the degradation of the Himalayas. The forested areas in the hills have considerably diminished in recent years. The rapid increase in population has led to a sharp decline in per capita agricultural and forested areas. As harvesting practices increase and open spaces and forests shrink, grazing activities have also intensified. Continuous and regular grazing has exposed the rock and soil layers, reducing soil compactness and leading to soil erosion. The unplanned growth of tourism has resulted in environmental deterioration and disruption to the ecosystem, which, in turn, has affected tourist traffic in the area. While tourism is undoubtedly a part of a nation's socioeconomic development, from a sustainable development perspective, this industry has yielded many negative consequences.

Regrettably, the rural economy of Darjeeling is highly unsatisfactory. It is noticeable that Darjeeling does not feature on the industrial map of the country. The economic viability and environmental suitability of the region do not favour the establishment of industries. However, despite the significant potential for small and cottage industries in the area, they have not been adequately introduced. The history of the region bears witness to the neglect faced by its people.

The persistent issues of poverty and lack of education have become deeply ingrained over time. Moreover, the infrastructure in the region is severely underdeveloped. Merely observing the towns and roadside developments in the Queen of the Hills cannot provide an accurate analysis of the economic structure. To truly understand the challenges faced by the people, a comprehensive survey of rural villages is necessary. Such a survey would reveal the harsh reality of how people struggle to make a living. The unsatisfactory state of infrastructure is evident when observing the residents of remote villages who have to endure a day-long journey on foot to access shopping facilities in towns. This indicates a lack of proper transportation and communication facilities in many villages. The rural population has to travel long distances on foot to sell their produce in the nearest town. During the rainy season, the situation worsens as frequent large-scale landslides and other forms of mass wasting occur in the hills. This leads to a significant loss of life among villagers. Additionally, access to safe drinking water, educational institutions, primary healthcare centres, and reliable power supply face similar challenges in rural areas of the hills.

Furthermore, rural areas in Darjeeling have experienced a noticeable trend of male-selective migration in search of better opportunities in the past few decades. This migration primarily occurs towards urban environments within the district and major cities such as Delhi, Kolkata, Chennai, and Mumbai, to name a few important destinations. Several factors have contributed significantly to this phenomenon, including population pressure, degradation of common property re*Source*s like forests, long neglect of the region, economic and social backwardness, low agricultural productivity, and lack of opportunities. While this migration brings in much-needed income, it also diminishes the available workforce in the home villages. As a result, agricultural production has been severely disrupted and is ultimately declining. The outflow of labour force not only damages the subsistence economy but also discourages efforts towards self-reliance in the villages. Although the money earned through migration is sent back to the villages to purchase provisions, ecological constraints, inadequate manpower, and reliance on traditional low-tech agriculture prevent the local hill population from generating sufficient income to meet their basic needs. Consequently, the money circulates back to the plains to purchase provisions transported from there, creating a vicious cycle that leaves the rural people of Darjeeling Himalaya with minimal savings.

3.3.6 Polity

For over half a century, the Queen of Hills has witnessed a prolonged agitation for the establishment of a separate state called "Gorkhaland" outside the boundaries of West Bengal but within the Indian Union. This historic movement, which took place in the early 1980s, had a profound impact on the state of West Bengal. The movement was led by the Gorkha National Liberation Front (GNLF), which was formed in July 1980. It quickly gained momentum and became a mass movement representing the Nepali-speaking population in the three hill subdivisions of the district. The agitation had several key demands, which included:

- The creation of a separate state of Gorkhaland that would be free from the administrative control of West Bengal.
- The repeal of clause VII of the Indo-Nepal Treaty of 1950.
- Recognition and inclusion of the Nepali language in the eighth schedule of the Indian constitution.
- Employment opportunities and economic development for the local people and the region respectively.

SubashGhising, a former army personnel, poet, and novelist, played a leading role in the Gorkha National Liberation Front (GNLF). However, the pursuit of autonomy has been a central theme in Darjeeling politics since before India's independence. Various local political groups had raised demands for autonomy in 1907, 1917, 1920, and 1934. It was during the period of 1980-85 that the GNLF truly championed the cause for a separate statehood for Darjeeling, including the adjacent areas of Dooars in Jalpaiguri, where over 50% of the population consisted of Nepali inhabitants. In 1980, the issue was invoked for the first time by referring to Article 3(a) of the Indian Constitution.

The demand for the abrogation of Article 7 of the Indo-Nepal Treaty of 1950 and the question of citizenship for Indian Nepalis emerged later on. Article 7 allowed

Nepalis from Nepal to cross over to India, reside, own property, and engage in trade and commerce. The GNLF claimed that this clause posed a threat to the status of Indian Nepalis, as they would be mistakenly identified with Nepalese nationals from Nepal and could face eviction. To emphasize the distinction between Indian Nepalis and Nepalese Nepalis, the terms "Gorkha" and "GorkhaBhasha" were increasingly used in place of Nepali and Nepali language, respectively.

It was widely believed that the residents of the Darjeeling district were distinct from West Bengal in terms of ethnicity, culture, language, social structure, and historical background. As a result, there was an increasing demand for constitutional recognition of the Nepali language, particularly since 1972, in order to preserve the ethnic and linguistic identity of the hill people. Despite Morarji Desai openly dismissing the demand and labeling Nepali as a foreign language, it was during Rajiv Gandhi's tenure that the central government recognized the "Gorkhas" as Indian citizens through a gazette notification. In the early 1990s, the Nepali language was eventually included in the eighth Schedule of the Indian Constitution, although it was not officially recognized as GorkhaliBhasa. Furthermore, the Gorkhaland Movement gained momentum due to various factors, including rising unemployment, inadequate infrastructure, limited educational and medical facilities, increasing deforestation, lack of opportunities, and most significantly, the perceived neglect and unfavourable treatment of the hill people by the government of West Bengal. These factors contributed significantly to the impetus behind the Gorkhaland Movement.

Between 1986 and 1988, a series of violent incidents and counter-violence occurred in the hills of Darjeeling, parts of Terai, and Dooars in the Jalpaiguri district. In response to the violence, the central government deployed forces such as CRPF, BSF, and state Armed Police. The Gorkha National Liberation Front (GNLF), being a political organization, had its armed wing called the Gorkha Volunteer Corps (GVC). However, the GVC maintained a separate identity within the GNLF. By the end of 1986, GNLF cadres began arming themselves, with the exception of the Kalimpong unit of the GVC, where reconciliation did not take place. Consequently, there were significant clashes between the GNLF and GVC in Kalimpong. In 1988, a former army personnel named ChhatreSubba emerged as the leader of the GVC and the president of the Gorkha Liberation Organisation (GLO) from Kalimpong itself. The clashes between the followers of GNLF and GVC were characterized by frequent kidnappings and raids on each other's strongholds. Many individuals were killed, beheaded, and subjected to torture during these clashes. Intra-GNLF clashes also occurred, primarily due to the high-handedness of SubashGhising. Mutual mistrust and enmity existed between nationalist factions and those who had initially supported Marxism but had now turned nationalist. By the end of 1987, over 500 houses belonging to Marxist Nepalis were set on fire, and more than 65 Marxists were killed. It is argued that these fratricidal clashes could have been avoided if the GNLF had focused solely on fighting for ethnic and linguistic identity rather than engaging in internal power struggles.

By the mid-1988, the agitation for Gorkhaland came to a halt, and on August 22, 1988, the leader of the GNLF, Mr.Ghishing, signed a political document in the presence of Union Home Minister Mr.Buta Singh and West Bengal Chief Minister Mr.JyotiBasu. In this document, Mr.Ghishing made a promise not to pursue the demand for Gorkhaland any further. In return, the Darjeeling hills were granted an autonomous council known as the Darjeeling Gorkha Hill Council (DGHC), which consisted of the three hill sub-divisions and aimed to address the interests of the hill people. The DGHC was given independent financial authority to utilize development funds for these areas, with the approval of both the central and state governments. However, a majority of the local population were dissatisfied with this decision and expressed their opposition, but their voices were suppressed.

During the 1990s, significant changes occurred in the political landscape of Darjeeling. The DGHC was composed of elected councillors who were granted autonomy to function as an independent body. However, as time passed, the councillors became overconfident and complacent in their roles. The ease of winning elections led to a lack of motivation to work effectively. Additionally, many of the councillors were illiterate or uneducated, which resulted in frequent mismanagement of funds. Observers note that Mr.Ghishing repeatedly failed to fulfill the expectations of the people.

The Gorkha National Liberation Front (GNLF) experienced a decline in its regional dominance in 2007 and 2008, as a new party called the GorkhaJanmuktiMorcha (GJM) emerged. The GNLF had previously led a violent movement for Gorkhaland in the 1980s and subsequently governed an autonomous council in the region. However, during its tenure, the GNLF became increasingly viewed as authoritarian and corrupt.

In October 2007, the GJM capitalized on the loss of faith in the GNLF's commitment and capability to achieve Gorkhaland. It successfully positioned itself as the new leader of the agitation, recognized by the government. To gain the trust of the public, GJM president BimalGurung, a former GNLF leader, actively distanced himself and the new party from what he referred to as "kutniti," a term used in Darjeeling to describe corrupt and dirty politics. Gurung emphasized the unity of the Gorkha community beyond political affiliations and urged people to unite under the GJM's banner. He accused the GNLF of betraying the people and selling the land of Darjeeling for personal gain. Gurung took oaths on various religious books, pledging to never betray the cause of the Gorkhas. By equating the GNLF with "dirty politics," the GJM utilized an anti-political rhetoric to position itself as a clean, honest, and dedicated proponent of the people's aspirations. The GJM presented itself as leading a genuine people's movement for Gorkhaland.

Frequent shutdowns, known as bandhs, have become a recurring occurrence, regular political turmoil, dirty politics and their impact is starting to be felt on the economy. In summary, the current state of Darjeeling Himalaya can be described as a state of confusion. It is now in the hands of the local people to determine the path they want to take for the future of the "Queen of the Hills."

3.4 Mental Modeling

Mental modeling refers to the cognitive process of creating mental representations or models of how things work or how different elements relate to each other. It involves constructing internal frameworks, concepts, or schemas that help individuals understand, interpret, and make sense of the world around them. Mental models can be based on past experiences, knowledge, beliefs, and assumptions, and they play a crucial role in perception, decision-making, problem-solving, and learning.

Mental modeling allows individuals to simplify complex information, fill in gaps, and make predictions about the future based on their understanding of causal relationships and patterns. These mental representations can be dynamic and adaptable, continuously updated as new information is acquired. Mental models can also influence behavior by guiding actions and expectations. In various fields such as psychology, cognitive science, and human-computer interaction, mental modeling is studied to gain insights into how individuals perceive, interpret, and interact with the world. It helps researchers understand how people form beliefs, make decisions, and solve problems based on their internal representations and mental simulations. Mental modeling can also be applied in fields like education, design, and communication to enhance learning, improve user experience, and facilitate effective communication by aligning mental models between individuals.

3.4.1 Mental Models

Mental models are cognitive representations that aid our comprehension of intricate concepts and facilitate improved decision-making. They serve as frameworks for thinking and problem-solving, enabling us to approach problems from various perspectives, generate innovative solutions, and enhance our cognitive abilities. Mental models are constructed based on past experiences, beliefs, and assumptions, allowing us to grasp the workings of the world around us. These models can be conscious or unconscious, varying in accuracy and usefulness depending on the specific context. Mental models play a crucial role in decision-making, problem-solving, and the acquisition of knowledge. They also contribute to effective communication and collaboration within group settings. However, it is important to note that mental models can introduce biases and errors if they are incomplete, inaccurate, inflexible, or resistant to change. In everyday life, we can employ mental models to better understand our environment and make more informed decisions.

Problem-solving is one area where mental models are frequently utilized. For instance, the "5 Whys" technique can help us comprehend a user's thought process and diagnose the underlying cause of a problem by asking a series of "why" questions.

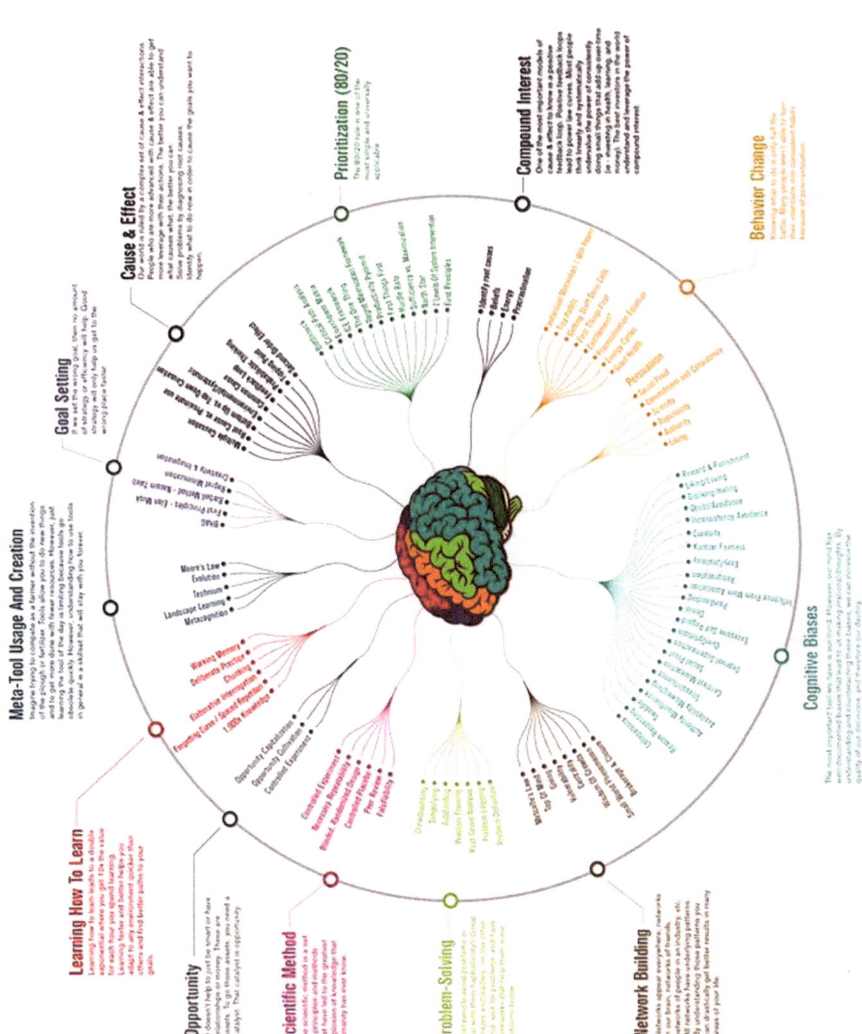

Fig. 3.4.1.1: Mental Models for decision-making

When it comes to decision-making, mental models aid in assessing the potential outcomes of different choices and determining which option is most likely to yield desirable results. An example of this is the "cost-benefit analysis," which evaluates the costs and benefits associated with various options, taking into account financial, social, or environmental impacts.

Critical thinking methodologies, such as the scientific or Socratic methods, encourage individuals to question their assumptions and challenge widely accepted beliefs. These approaches foster a deeper understanding and examination of our mental models.

3.4.2 Principles of Mental models

1. **Mental models encompass various possibilities:** each model encompasses a unique set of possibilities related to the current description.
2. **Mental models are visual representations**: the structure of a model corresponds to the structure of the represented concept or phenomenon (Peirce, 1931–1958, Vol. 4). For example, kinematic models depict a temporal sequence of events unfolding over time (Khemlani et al., 2013). However, models can also incorporate abstract symbols, such as a symbol for negation (Khemlani et al., 2012).
3. **The truth-based principle:** Mental models exclusively represent what is true in each possibility, omitting what is false. They facilitate rapid intuitive insights. In contrast, fully explicit models represent both true and false aspects, but constructing them requires more deliberation and access to working memory.
    - **Simplification:** Mental models simplify complex concepts and phenomena by capturing the essential elements and relationships. They help us understand and navigate the world by breaking down information into manageable chunks.
    - **Abstraction:** Mental models involve abstracting and generalizing information to create conceptual frameworks. They focus on key patterns and principles rather than specific details, allowing us to apply our understanding to different situations.
    - **Causality:** Mental models emphasize cause-and-effect relationships. They help us identify the factors that influence outcomes and predict how changes in variables can impact the overall system.
    - **Visualization**: Mental models often involve visual or spatial representations. By visualizing concepts, processes, or relationships, we can enhance our understanding and memory retention.
    - **Dynamic Nature:** Mental models are not static but evolve and adapt over time. They are updated based on new information and experiences,

allowing us to refine our understanding and improve decision-making.

- **Holistic Perspective:** Mental models aim to capture the interconnectedness and interdependencies within systems. They encourage a holistic view that considers multiple factors and their interactions rather than isolating individual components.
- **Bias Awareness:** Mental models can be influenced by biases and assumptions. Being aware of these biases helps us critically evaluate our models and consider alternative perspectives to avoid errors in thinking.
- **Learning and Feedback:** Mental models are refined through continuous learning and feedback. Testing and adjusting our models based on real-world observations and outcomes enable us to improve their accuracy and effectiveness.
- **Flexibility:** Effective mental models are flexible and adaptable. They allow for the incorporation of new information and the ability to adjust strategies and approaches as needed.
- **Communication and Collaboration:** Mental models facilitate effective communication and collaboration by providing a shared understanding of concepts and facilitating the exchange of ideas. Aligning mental models among individuals promotes clearer communication and better collaboration within teams or groups.

3.4.3 Reasoning with Mental models

The theory of mental models suggests that deductive reasoning can be understood as a three-step process. The first step involves comprehending the premises and constructing a mental representation, or mental model, of the described situation. According to this theory, the mental model reflects the logical structure of the situation. When multiple premises are involved, coordination heuristics are employed to integrate the information provided by each premise.

In the second step, a potential conclusion is generated based on the resulting mental model. An item of information is considered a conclusion if it is both consistent across all constructed models and not explicitly stated in the premises. These first two steps primarily involve comprehension and do not encompass a distinct reasoning process.

In contrast, the third step involves the subject attempting to falsify the reached conclusion by constructing alternative models of the premises. If no alternative model can be constructed in which the conclusion does not hold, then the conclusion is considered necessarily true. However, if an alternative model can be constructed in which the conclusion does not hold, the conclusion is deemed uncertain.

The primary prediction of the theory is that reasoning tasks that involve a larger number of mental models should be more challenging. This increased difficulty can manifest as higher error rates, perceived difficulty by the subject, and longer completion times. The reason for this difficulty is attributed to the limitations of working memory, which is responsible for both constructing and manipulating mental models. To cope with this cognitive load, individuals are expected to reduce the load by representing some information implicitly. This leads to the construction of simplified initial models. If necessary based on the specific task, these initial models can be expanded or elaborated to create a comprehensive representation of the described situation.

Evans (1993) proposed modifications to the mental models theory to account for a specific phenomenon, suggesting two explanatory factors. The first factor is a negative conclusion bias, which leads to a higher acceptance rate of negative conclusions compared to affirmative conclusions. This bias favours the acceptance of conclusions in the form of denial affirmatives (DA), where the negative conclusion "not q" is accepted, while hindering the acceptance of affirmative conclusions (AC) when both the antecedent and consequent of the major premise ("if p then q") are affirmative propositions.

The second factor is the directional nature of models for conditionals, which are oriented from the antecedent to the consequent. According to this proposal, Modus Ponens (MP) and Denying the Antecedent (DA) are considered forward inferences, making them easier to endorse than the corresponding backward inferences of Affirming the Consequent (AC) and Modus Tollens (MT). For the first model, AC involves a minor premise "q" as a backward inference, while for the second model, MT involves a minor premise "not q." The directionality of the models, as described by Ormerod, Manktelow, and Jones (1993), contributes to these differences.

As a result, both the negative conclusion bias and the directionality of models favour the acceptance of DA (a forward inference leading to a negative conclusion) and hinder the acceptance of AC (a backward inference with a positive conclusion). This leads to equal frequencies of endorsement, even though DA requires an additional model compared to AC.

3.4.4 Mental Model of Dynamic Systems

The mental model of dynamic systems refers to the cognitive representation individuals construct to understand and reason about complex, interconnected systems that involve various elements and their interactions over time. This mental model allows individuals to comprehend the behaviour, patterns, and dynamics of these systems.

Here are key aspects of the mental model of dynamic systems:

1. **System Elements:** The mental model includes an understanding of the individual elements or components that make up the system. These elements can be tangible objects, variables, or entities that influence the system's behaviour.

2. **Relationships and Interactions:** The mental model encompasses the relationships and interactions between the different elements of the system. It captures how changes in one element can affect others and how feedback loops or causal loops contribute to the overall system dynamics.
3. **Feedback Loops:** The mental model incorporates the concept of feedback loops, which are recurring patterns of interactions within the system. These loops can be reinforcing (positive feedback) or balancing (negative feedback) and have a significant impact on the system's stability and behaviour over time.
4. **Cause-and-Effect Relationships:** The mental model includes an understanding of cause-and-effect relationships within the system. It enables individuals to recognize how changes in certain elements or variables can lead to specific outcomes or consequences within the system.
5. **Time and Dynamics:** The mental model recognizes the importance of time in understanding dynamic systems. It accounts for the changes and evolving patterns that occur over time, allowing individuals to perceive trends, delays, and non-linear behaviours within the system.
6. **Visualization:** Visualization plays a crucial role in the mental model of dynamic systems. Individuals often use diagrams, flowcharts, or mental imagery to represent and visualize the structure, connections, and flows within the system.
7. **Complexity and Emergence:** The mental model acknowledges the complexity of dynamic systems and the potential for emergent properties or behaviours to arise from the interactions between elements. It allows individuals to grasp the system's overall behaviour beyond the sum of its individual parts.
8. **Updating and Adaptation:** The mental model of dynamic systems is not static but is continually updated and adapted based on new information and experiences. Individuals refine their mental models as they gain insights into the system's behaviour and make predictions about future states or scenarios.

3.4.5 Mental model in relation to system dynamics and systemic thinking

In relation to system dynamics and systems thinking, a mental model serves as an individual's internal representation of a complex system and its dynamics. It encompasses their understanding of the interrelationships, feedback loops, cause-and-effect relationships, and overall behaviour of the system.

System dynamics: System dynamics is an interdisciplinary approach to understanding and analysing complex systems. It involves studying the behaviour of systems over time by focusing on the feedback loops, interrelationships between components, and the flow of information and re*Source*s within the system.

Developed by Jay W. Forrester in the 1950s, system dynamics combines principles from mathematics, computer simulation, and social sciences to model and simulate dynamic systems.

Systemic thinking: systemic thinking, also known as systems thinking, is an approach to understanding and addressing complex problems by considering the interconnectedness and interdependencies of the various components within a system. It involves analysing the system as a whole rather than focusing on isolated parts or individual events.

Here's how a mental model relates to system dynamics and systems thinking:

1. **System Understanding**: A mental model allows individuals to comprehend the structure and functioning of a system. It helps them identify the key components, their interactions, and the overall behaviour of the system over time. This understanding is crucial for system dynamics, which focuses on analysing how variables and feedback loops influence the behaviour of complex systems.
2. **Feedback Loops:** Mental models capture the feedback loops within a system, including reinforcing (positive) and balancing (negative) feedback loops. Individuals with a strong mental model can identify how changes in one aspect of the system can impact other components and how feedback mechanisms shape the system's behaviour. This understanding is fundamental to system dynamics, as it emphasizes the role of feedback in driving system behaviour.
3. **Causal Relationships**: Mental models help individuals recognize cause-and-effect relationships within a system. They enable individuals to understand how changes in variables or components can lead to specific outcomes or consequences. This comprehension is vital for systems thinking, which involves analysing the underlying causes and understanding the systemic nature of problems and solutions.
4. **System Behaviour over Time:** A mental model enables individuals to grasp the dynamic nature of a system. It incorporates an understanding of how the system evolves and behaves over time, including trends, delays, and nonlinear patterns. This temporal perspective is a key aspect of both system dynamics and systems thinking.
5. **Holistic Perspective:** Mental models encourage a holistic perspective on systems. They help individuals consider the interdependencies and interactions between various components rather than viewing them in isolation. This holistic thinking is central to systems thinking, which emphasizes the interconnectedness and complexity of systems.
6. **Simulation and Analysis**: Mental models can be used to simulate and analyse system behaviour. By manipulating variables within their mental

models, individuals can explore different scenarios and assess the potential impacts on the system. This simulation and analysis align with the modeling and experimentation practices of system dynamics and systems thinking.

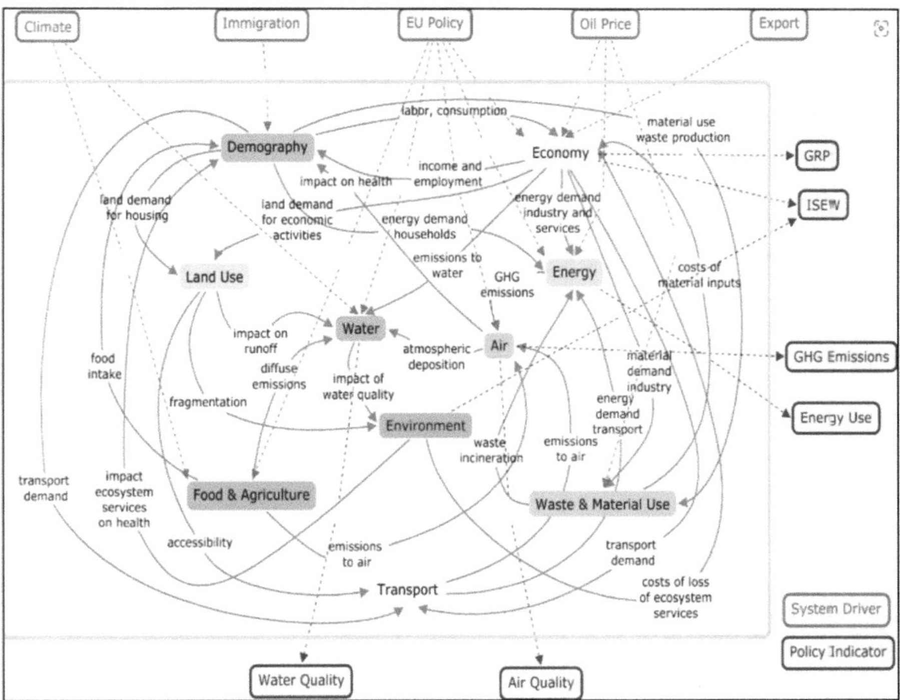

Fig. 3.4.5.1: System Dynamics modelling

4

Social Ecology

4.1 Introduction

The research setting pertains to the comprehensive details of a specific area's social environment where the study is carried out. The study area typically encompasses a distinct geographic region such as a state, district, block, or gram panchayat area, chosen based on the convenience of the researcher who possesses sufficient knowledge about the locality, including its communication infrastructure. This ensures easy access to every part of the area for collecting data.

Migration is a complex phenomenon that has significant implications for individuals, communities, and countries worldwide. It involves the movement of people from one geographical location to another, either within a country or across international borders. The research setting of migration refers to the specific context in which migration patterns and processes are studied. Understanding the research setting is crucial for comprehending the underlying factors, dynamics, and consequences of migration.

4.2 Profile of the State

4.2.1 General Information of the State

West Bengal is a state located in the eastern part of India, situated along the Bay of Bengal. It has a population of over 91 million people and covers an area of 88,752 square kilometres (34,267 square miles). West Bengal ranks as the fourth-most populous and thirteenth-largest state in India in terms of area, and it is the eighth-most populous country subdivision in the world. The Bengal region, which is a part of the Indian subcontinent, shares borders with Bangladesh to the east, Nepal to the north, and Bhutan to the northwest. It also shares boundaries with the Indian states of Jharkhand, Odisha, Bihar, Sikkim, and Assam. The state capital is Kolkata, which is the third-largest metropolitan city and the seventh-largest city in India by population. West Bengal encompasses diverse geographical features, including the Darjeeling Himalayan hill region, the Ganges delta, the Rarh region, the coastal Sundarbans, and the Bay of Bengal. The dominant ethnic group in the state is the Bengalis, with Bengali Hindus forming the majority. West Bengal is divided into 23 districts, namely Alipurduar, Coochbehar, Darjeeling, Jalpaiguri, Uttar Dinajpur, DakshinDinajpur, Malda, Murshidabad, Hooghly, Howrah, Nadia, Birbhum, PurbaBardhaman, PaschimBardhaman, Bankura, Purulia, North 24

Parganas, South 24 Parganas, PurbaMedinipur, PaschimMedinipur, Jhargram, Kalimpong, and Kolkata. The state is further divided into more than 341 Community Development Blocks, 3354 Gram Panchayats, and has over 42,000 villages.

Total area of the State	88,752 sq. km
Total population of the State	9,13,47,736
Total no of Districts	23
Total no of CD Blocks	341
Total no of Statutory Towns	129
Total no of Census Towns	780
Total no of Villages	40203

Source: West Bengal, Census of India(2011)

4.2.2 Geography and Climate of the State

West Bengal is located in the eastern part of India, stretching from the Himalayas in the north to the Bay of Bengal in the south. The state covers an area of 88,752 square kilometres (34,267 square miles). In the northern region of the state lies the Darjeeling Himalayan hill area, which is part of the eastern Himalayas mountain range. This area includes the highest peak in the state, Sandakphu, reaching an elevation of 3,636 meters (11,929 feet). The Terai region, a narrow strip, separates the hills from the North Bengal plains. Moving further south, the landscape transitions into the Ganges delta. The Rarh region lies between the Ganges delta in the east and the western plateau and highlands. In the extreme south, there is a small coastal region, and the Sundarbans mangrove forests are a notable geographical feature at the Ganges delta.

West Bengal experiences a varied climate, ranging from tropical savannah in the southern parts to humid subtropical in the north. The main seasons are summer, rainy season, a short autumn, and winter. The delta region is known for its high humidity during the summer, while the western highlands have a dry summer similar to northern India. Daytime temperatures can reach as high as 38 °C (100 °F) to 45 °C (113 °F). At night, a cool southerly breeze carries moisture from the Bay of Bengal. Monsoons bring rainfall to the entire state from June to September, with areas like Darjeeling, Jalpaiguri, and Cooch Behar experiencing heavy rainfall of over 250 centimetres (98 inches). The Bay of Bengal region often experiences storms due to low pressure during the monsoon season. Winter, occurring in December and January, is mild in the plains with average minimum temperatures around 15 °C (59 °F). A cold and dry northern wind blows during winter, resulting in lower humidity levels. The Darjeeling Himalayan hill region, on the other hand, experiences a harsh winter with occasional snowfall.

4.2.3 Agriculture of the State

Rice and potatoes are recognized as the primary food crops in West Bengal. The state holds the largest share in rice production, a staple food across India, with an annual output of approximately 16.1 million tonnes in the fiscal year 2015-2016. In the fiscal year 2017-2018, rice production in the state reached 14.99 million tonnes. West Bengal is the second-largest producer of potatoes in India, with an average annual output of 11 million tonnes in fiscal year 2015. Additionally, it is the second-largest fish-producing state in the country.

Other significant crops in West Bengal include sugarcane and oilseeds. Jute cultivation is particularly prominent along the border with Bangladesh and in the southern regions south of the Ganges River. The southern and central parts of the state are known for the widespread production of mangoes, jackfruit, and bananas. Winter crops such as wheat and potatoes are cultivated throughout the southern areas. The northern regions, including Darjeeling and Jalpaiguri districts, are renowned for their high-quality tea production. West Bengal is the second-largest tea-producing state in India and is internationally recognized for the premium Darjeeling tea variety. Moreover, the Darjeeling region also cultivates oranges, apples, pineapples, ginger, and cardamom.

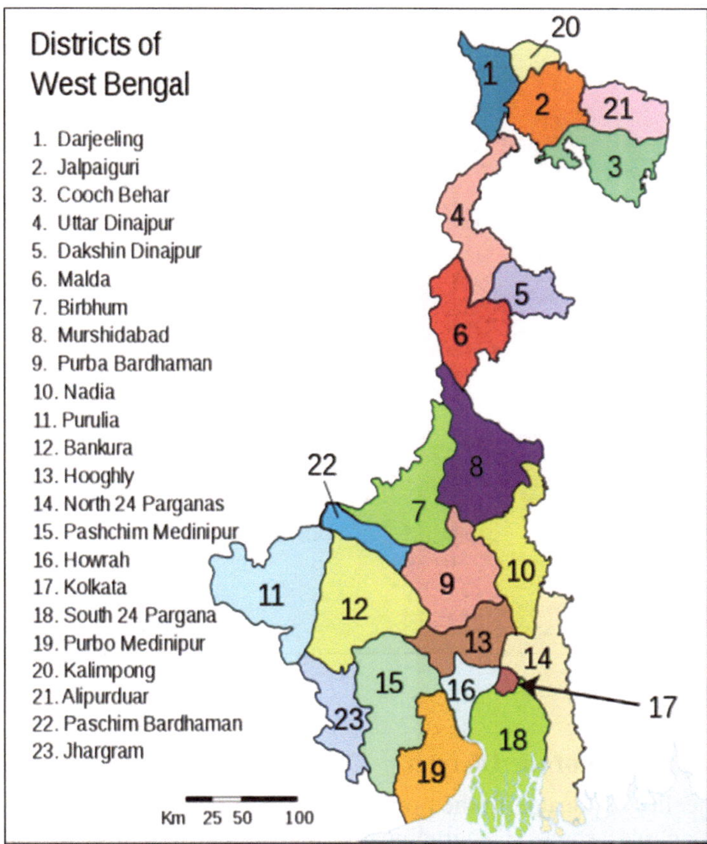

Fig. 4.2: Districts Map of West Bengal

Fig. 4.2.1: Map of West Bengal

4.3 Profile of the District:

4.3.1 General Information of the District

The district of Darjeeling is located in the northernmost part of West Bengal and has an irregular triangular shape. Geographically, the northern portion of the district is part of the Great Himalayas, ranging in elevation from 300 to 12,000 feet above sea level. The southern portion of the district consists of the Tarai region, which stretches along the base of the hills. Situated in the Eastern Himalayas, Darjeeling has an average elevation of 2,045 meters (6,709 feet).

To the west of Darjeeling lies the easternmost province of Nepal, while the Kingdom of Bhutan is situated to the east. The Indian state of Sikkim borders the district to the north, and farther north lies the Tibet Autonomous Region of China. Bangladesh is located to the south and southeast, while the majority of West Bengal lies to the south and southwest, connected to the Darjeeling region by a narrow tract. On clear days, the world's third-highest mountain, Kangchenjunga, can be prominently seen in the north.The name "Darjeeling" is derived from the Tibetan words "dorje," meaning thunderbolt (originally the sceptre of Indra), and "ling," which refers to a place or land. Thus, Darjeeling translates to "the land of the thunderbolt." The history of Darjeeling dates back before its acquisition by the East India Company in 1835. It was previously part of Sikkim and briefly belonged to Nepal.

The Darjeeling Himalayan Railway, also known as the Toy Train, gained recognition as a World Heritage Site by UNESCO following an international campaign in its support. In 2005, Darjeeling tea was granted a geographical indication (GI) by the World Trade Organization, offering protection to the brand and contributing to the region's development.

Darjeeling District can be categorized into Hill and Plain regions based on its physiographical features. Among the 12 Community Development Blocks and 5 Statutory Towns in the district, 8 CD Blocks (Darjeeling-Pulbazar, RangliRangliot, JorebungalowSukhiapokhri, Kalimpong I, Kalimpong II, Gorubathan, Mirik, and Kurseong) and 4 Statutory Towns (Darjeeling Municipality, Kurseong Municipality, Kalimpong Municipality, and Mirik Notified Area) are situated in the hilly areas, while the remaining 4 CD Blocks (Matigara, Naxalbari, Phansidewa, and Kharibari) and Siliguri Municipal Corporation are located in the plains.The district comprises a total of 687 villages and 24 Census Towns. In terms of population,

Darjeeling District ranks 18th in both total population and child population (0-6 years) within the state. It holds the 18th position for Scheduled Castes population and the 5th position for Scheduled Tribes population in the state. The urban population share of Darjeeling District has increased from 32.3% in the 2001 Census to 39.4% in the 2011 Census. The district ranks 8th in terms of decadal population growth rate with a rate of 14.8% among all districts.

Due to its hilly topography, a significant portion of Darjeeling District has sparse population density, ranking 17th in the state with a density of 586 people per square kilometre according to the 2011 Census. Notably, the district has the highest sex ratio (970 females per 1000 males) in the state, and it retains the same rank for rural sex ratio (973). In terms of Scheduled Caste and Scheduled Tribe population, it ranks 3rd (965) and 2nd (1015) respectively. In relation to the proportion of child population, Darjeeling District ranks 13th in the state at 10.5%.

However, there has been a decline in the child sex ratio (CSR) in the district, dropping from 962 in the 2001 Census to 953 in the 2011 Census, similar to the overall decline in the state from 960 to 956. The district covers an area of 3149.00 square kilometres, ranking 14th in size jointly with Hugli District. The literacy rate in Darjeeling District is 79.6%, higher than the state average of 76.3%, placing it in the 6th position in the state. The Work Participation Rate has also improved from 35.4% in the 2001 Census to 37.0% in the 2011 Census, earning the district the 13th rank in this aspect.

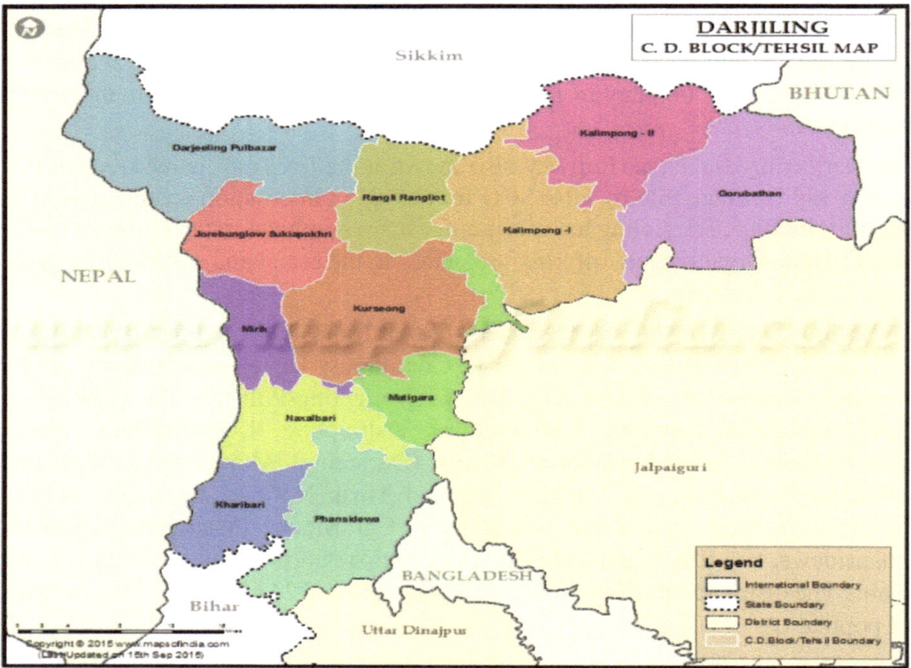

Fig. 4.3: Tehsil Map of Darjeeling

4.3.2 Geography and Climate of the District

The Darjeeling hills, also known as the Darjeeling Himalayan hill region, consist of parts of the Darjeeling district and the entire Kalimpong district. Specifically, it includes the subdivisions of Darjeeling Sadar, Kalimpong, and Kurseong. The town of Darjeeling is situated in the Sadar subdivision at an average elevation of 2,045 meters (6,709 feet). These hills are located on the Darjeeling-Jalapahar range, which runs from south to north, starting at Ghum. The range has a Y-shaped structure, with its base resting at Katapahar and Jalapahar, and two arms diverging northwards from the Observatory Hill. The northeastern arm descends rapidly and terminates in the Lebong spur, while the northwestern arm slopes gently, passing through North Point, and ends in the valley near the Tukver Tea Estate. One of the most prominent mountains visible from this region is Kangchenjunga, the third-highest peak in the world, reaching a height of 8,598 meters (28,209 feet). It is located approximately 74.4 kilometers (46.2 miles) to the north.

The formation of the Darjeeling hills is a result of the accumulation of folds, faults, and tangential thrusts caused by compression in the north-south direction. This compression occurs as the Indian tectonic plate subducts beneath the Eurasian plate. The physical composition of the hills varies across different regions. In the southern areas, there are unaltered sedimentary rocks, while the middle and northern regions contain various types of metamorphic rock and some intrusive rocks. This suggests that there has been an upward intrusion of the Earth's mantle. Over time, the collective process of shearing, folding, crushing, and fracturing has weakened the rocks, making them susceptible to water infiltration through crevices. This water percolation leads to the accumulation of pore water pressure. The hills around Kalimpong, located to the east, consist of phyllites and schists, whereas gneiss is predominant in the western regions, including Darjeeling.

The Teestariver, which is the primary river in the Darjeeling area, originates from a glacier in Sikkim at an altitude of 6,300 meters (20,700 ft.). It travels in a southern direction, initially converging with the Rangporiver and then the Rangeet river, before leaving the hilly region and ultimately merging with the Brahmaputra river in Bangladesh. During the summer monsoon, the Teestariver has a flow rate of 1,500 cubic meters (53,000 cu ft.) per second.

Darjeeling experiences a temperate subtropical highland climate, classified as Cwb according to the Köppen climate classification system. The average yearly precipitation in Darjeeling is around 3,100 mm (120 in). The majority of the rainfall, about 80%, occurs between June and September due to the monsoon season in South Asia. The amount of rainfall plays a very important role in causing instability of slopes. A very high intensity of rainfall within a short span of time is not uncommon in Darjeeling hill areas. It is found in the old records; that this natural phenomenon has occurred about 42 times during the period from 1891 to 1975 (Chatterjee, 1982). There is a significant increase in rainfall from May to June, with a ratio of 2.6 or 260%. On the other hand, only 3% of the annual rainfall happens between December and March. Darjeeling's higher altitude compared to other regions in the Eastern Himalayas at the same latitude (27° N), such as the Assam hills, results in higher levels of UV radiation due to the rarified air. During the peak months of May, June, and July, the average monthly UV radiance in Darjeeling is approximately 4500 microwatts per square cm per day. This level is 50% higher than that of the Assam hills to the east, which are situated at an altitude of 170 meters (560 ft.).

Climate data for Darjeeling (1981–2010, extremes 1901–2012)													
Month	Jan	Feb	Mar	Apr	May	Jun	Jul	Aug	Sep	Oct	Nov	Dec	Year
Record high °C (°F)	19.0 (66.2)	19.2 (66.6)	24.0 (75.2)	27.0 (80.6)	25.7 (78.3)	27.7 (81.9)	28.0 (82.4)	28.5 (83.3)	27.5 (81.5)	26.0 (78.8)	24.5 (76.1)	20.0 (68.0)	28.5 (83.3)
Average high °C (°F)	10.7 (51.3)	12.4 (54.3)	15.6 (60.1)	18.5 (65.3)	19.3 (66.7)	19.8 (67.6)	19.6 (67.3)	20.0 (68.0)	19.8 (67.6)	19.5 (67.1)	17.1 (62.8)	14.0 (57.2)	17.2 (63.0)
Daily mean °C (°F)	6.1 (43.0)	7.7 (45.9)	10.6 (51.1)	13.7 (56.7)	14.9 (58.8)	16.3 (61.3)	16.5 (61.7)	16.7 (62.1)	16.1 (61.0)	15.0 (59.0)	11.7 (53.1)	8.9 (48.0)	12.9 (55.1)
Average low °C (°F)	1.5 (34.7)	2.9 (37.2)	5.7 (42.3)	8.8 (47.8)	10.6 (51.1)	12.8 (55.0)	13.4 (56.1)	13.4 (56.1)	12.4 (54.3)	10.5 (50.9)	6.3 (43.3)	3.8 (38.8)	8.5 (47.3)
Record low °C (°F)	-7.2 (19.0)	-6.4 (20.5)	-4.8 (23.4)	0.0 (32.0)	1.4 (34.5)	6.6 (43.9)	3.9 (39.0)	8.0 (46.4)	6.2 (43.2)	3.2 (37.8)	-4.4 (24.1)	-4.6 (23.7)	-7.2 (19.0)
Average rainfall mm (inches)	13.5 (0.53)	14.0 (0.55)	30.8 (1.21)	76.9 (3.03)	137.9 (5.43)	466.0 (18.35)	656.7 (25.85)	528.2 (20.80)	379.7 (14.95)	59.1 (2.33)	14.4 (0.57)	2.9 (0.11)	2,380 (93.70)
Average rainy days	1.1	1.5	2.8	6.8	10.5	18.8	22.9	21.7	14.9	2.9	0.6	0.7	105.3
Average relative humidity (%) (at 17:30 IST)	81	78	75	78	88	93	94	92	90	84	75	74	84
Mean daily sunshine hours	5.4	5.0	4.7	4.9	4.9	2.4	2.5	3.3	3.2	5.4	6.3	6.1	4.5
Average ultraviolet index	5	6	9	11	13	15	15	14	12	9	6	4	10

Fig. 4.3.2.1: Climate data for Darjeeling (1981-2010)

Source: India Meteorological Department, UV Index.

4.3.3 Agriculture of the District

Darjeeling, located in the Indian state of West Bengal, is famous for its tea industry and is well-known for producing high-quality tea leaves. However, agriculture in Darjeeling extends beyond tea cultivation and includes a diverse range of crops due to its favourable climatic conditions and geographical features.

1. *Tea Plantations*: The most significant agricultural activity in Darjeeling is the cultivation of tea. The region's cool climate, abundant rainfall, and hilly terrain create ideal conditions for growing tea bushes. Darjeeling tea is renowned for its distinctive flavour and is often referred to as the "Champagne of Teas." Tea gardens cover large areas of land in the region, and the tea industry plays a crucial role in the local economy.
2. *Horticulture*: Darjeeling is also known for its horticultural produce. The region's cool climate and fertile soil support the cultivation of various fruits and vegetables. Apples, pears, oranges, plums, apricots, and strawberries are among the fruits grown in Darjeeling. Vegetables such as potatoes, cabbages, cauliflowers, and beans are also cultivated.
3. *Rice Cultivation*: Paddy fields can be found in the lower elevations of Darjeeling. Farmers grow rice, both for consumption and commercial purposes. The rice fields are usually terraced to accommodate the hilly terrain and prevent soil erosion.

4. *Floriculture*: The favourable climate and scenic landscapes of Darjeeling have made it a suitable location for floriculture. The region produces a variety of flowers, including roses, orchids, gladioli, lilies, and dahlias. These flowers are cultivated both for local consumption and for export to other parts of the country.
5. *Cardamom Plantations*: Darjeeling's climate and terrain are also suitable for growing spices. Cardamom plantations can be found in certain areas of the region. Cardamom is a valuable spice and is highly sought after in the culinary industry.
6. *Livestock Farming*: Livestock farming, including dairy farming and poultry, is practiced in Darjeeling. Cattle, goats, and poultry are reared for milk, meat, and egg production, meeting the local demand for animal products.

It is important to note that the specific agricultural practices and crops may vary depending on the altitude, soil type, and microclimatic conditions within different parts of Darjeeling. The agriculture sector in Darjeeling contributes significantly to the local economy, provides employment opportunities, and plays a vital role in sustaining the region's agricultural heritage.

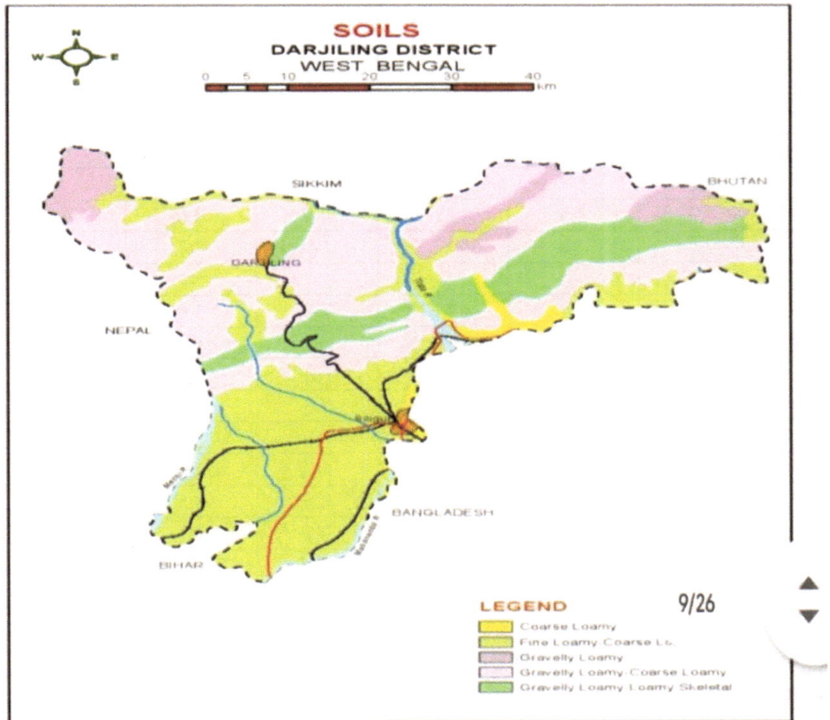

Fig. 4.3.3.1: Soil Profile Map of Darjeeling.

Source: NBSS & LUP Regional Centre, Kolkata

Table 4.3.3.1: Area under Principal Crops in the district of Darjeeling

(thousand hectares)

	Crops	2006-07	2007-08	2008-09	2009-10	2010-11
	(1)	(2)	(3)	(4)	(5)	(6)
Foodgrains						
1.	Rice	31.4	32.2	32.5	32.3	32.3
	Aus	4.4	3.8	3.6	3.5	3.5
	Aman	25.5	26.9	27.3	27.3	27.3
	Boro	1.5	1.5	1.6	1.5	1.5
2.	Wheat	2.2	2.2	2.0	2.0	2.0
3.	Barley	-	-	-	-	-
4.	Maize	14.6	15.3	15.8	15.8	16.1
5.	Other Cereals	12.0	11.8	11.7	11.8	11.7
	Total Cereals	60.2	61.5	62.0	61.9	62.1
6.	Gram	(a)	-	-	-	-
7.	Tur	(a)	(a)	(a)	(a)	(a)
8.	Other Pulses	2.0	1.1	1.2	1.1	1.1
	Total Pulses	2.0	1.1	1.2	1.1	1.1
	Total Foodgrains	62.2	62.6	63.2	63.0	63.2
Oil Seeds						
1.	Rapeseed & Mustard	0.2	0.2	0.2	0.2	0.2
2.	Linseed	0.1	0.1	0.1	0.1	0.1
3.	Other Oil seeds	14.0	0.3	0.3	0.2	0.2
	Total Oil seeds	14.3	0.6	0.6	0.5	0.5
Fibres						
1.	Jute	2.5	2.6	2.9	3.0	3.0
2.	Mesta	-	-	-	-	-
3.	Other Fibres	-	-	-	-	-
	Total Fibres	2.5	2.6	2.9	3.0	3.0
Miscellaneous crops						
1.	Sugarcane	-	(a)	(a)	(a)	(a)
2.	Potato	6.7	7.2	7.3	7.6	7.5
3.	Tobacco	-	-	-	-	-
4.	Tea	39.2	37.2	37.2	37.2	37.2
5.	Chillies (dry)	0.4	0.5	0.5	0.5	0.5
6.	Ginger	2.1	2.1	2.1	2.1	2.1
	Total Miscellaneous crops	48.4	47.0	47.1	47.4	47.3

(a) Less than 50 hectares

Sources : 1) Directorate of Agriculture, Govt. of W.B. 2) B.A.E.& S., Govt. of W.B.

4.3.4 Land Use Pattern of the District

The main problem in respect of land use in the Darjeeling Hill Areas is related to high density of population. There is very limited scope for extension of agricultural land to cope up with increasing pressure of population. As a result pressure on forested and other restricted areas is gradually increasing.

Another problem related to land use and consequent landslide is that in Darjeeling Hill Areas, roads have never been examined with its carrying capacity respect with geology etc. Along with new road construction the vehicular movements have increased to a great extent with the rapid growth of trade and commerce. Heavy traffic movements along with heavy rainfall are responsible for most of the landslide occurrences especially on the roads. In recent years, it has been observed that there is a constant increase in the vehicular traffic, especially heavy vehicles like trucks and buses. The record reflects that at present, the number of registered vehicles in the hill subdivisions are more than 6500.

The agricultural land use pattern of Darjeeling is primarily characterized and influenced by its diverse topography, climate, cultivation practices and the availability of suitable re*Source*s. Darjeeling is famous for its Tea plantations, but it also supports various other agricultural activities.

Multiple studies (Bairagya and Jha, 2015; Nad, 2015; Tirkey and Nepal, 2012) indicate that landslides have a severe impact on crop cultivation, particularly in the tea gardens of the steep slopes in the Darjeeling hill area. This vulnerability is amplified during earthquakes and heavy rainfall. Despite being located in the eastern Himalayas and the Himalayan Frontal Thrust, the presence of the Tista Lineament (which trends NW-SE) and the Kanchenjunga Lineament (which trends NE-SW) in Darjeeling indicates that the region is prone to earthquakes. It is observed that areas susceptible to seismic activity, especially those with tea gardens, experience more pronounced effects of landslides (Bairagya andJha, 2015).

Cbssification of Land Utilisation Statistics in th? district of Darjeeling

(Thousand hectares)

Year	Reporting Area	Forest Area	Area under Non-agncullural use	Barren 4 uncullura Ne land	Permanent pastures A other grazing land	Land under Mise free groves not included in Nel area sown	Cullutable waste land	Fallow land other than Current (allow	Current fallow	Nel area sown
(1)	(2)	(3)	(4)	(5)	(6)	(7)	(8)	(9).	(10)	(11)
2008-09	325.47	124.57	39.88	2.48	0.87	221	1.67	3.75	17.53	132.51
2009-10	325.47	124.57	40.16	2.14	1.13	2.33	1.55	3.65	17.67	132.27
2010-11	325.47	124.57	40.53	2.46	0.83	2.35	1.49	3.22	16.44	133.58
2011-12	325.47	124.57	38.62	2.57	0.57	2.64	1.31	3.18	17.36	134.65
2012-13	325.47	124.58	3921	2.33	0.50	2.78	1.12	2.93	16.55	135.47

Source &recterate(rfAgriciiture(Evafia6on).GoAo(W.B.)

Fig. 4.3.4.1: Classification of Land Utilisation Statistics of Darjeeling

Source: Directorate of Agriculture (Evaluation), Govt. of W.B.

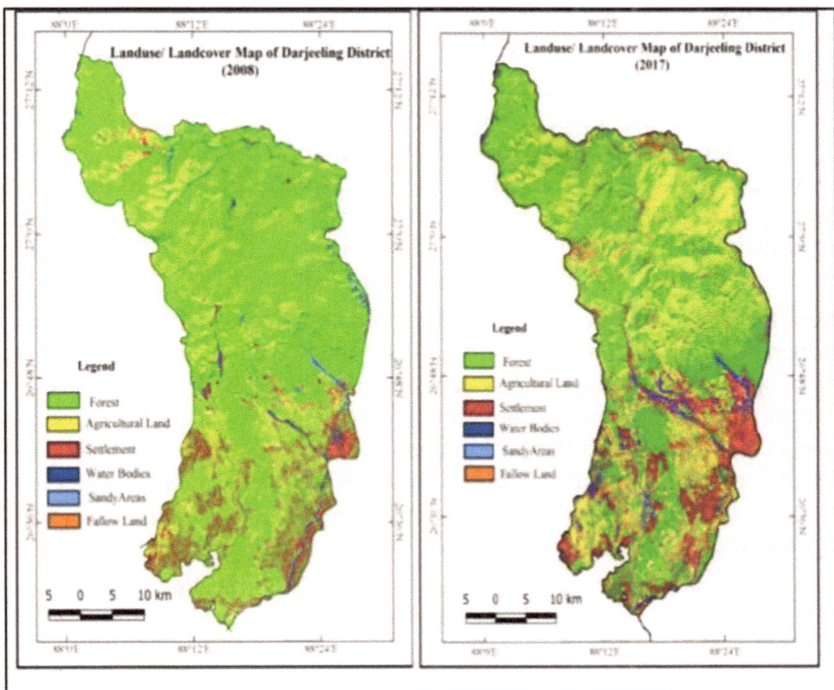

Fig. 4.3.4.2: Landuse/ Landcover maps of the study area of 2008 and 2017

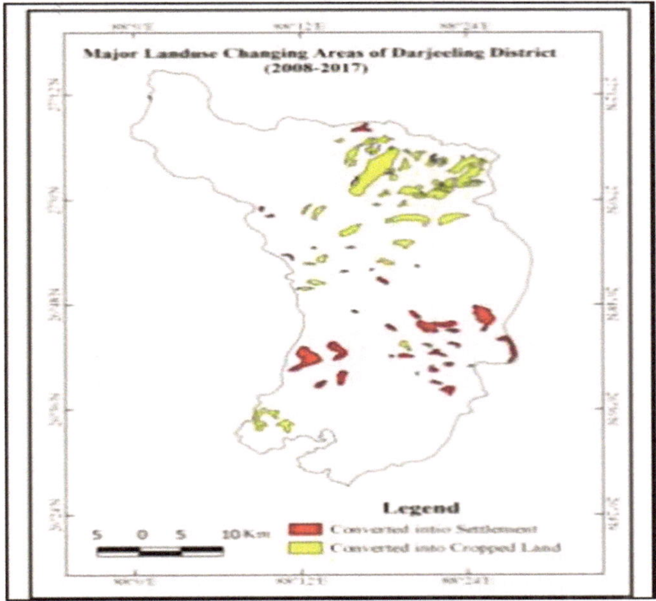

Fig. 4.3.4.3: Major Converted Areas from 2008 to 2017

4.3.5 Irrigation

The primary *Source* of water for the Darjeeling municipality is the Senchal Wildlife Sanctuary, situated about 11 kilometres southeast. Spanning an area of 37.97 square kilometres, the sanctuary's elevation ranges from 1,500 to 2,600 meters. The water supply mainly relies on natural springs within the sanctuary, although not all of them flow throughout the year. The steep slopes of the nearby ridges, with inclines between 20° and 48°, contribute to significant surface run-off, allowing water to be collected in partially confined spaces. When the groundwater volume reaches a critical level, it emerges as seasonal springs.

To harness the water re*Source*s, 26 perennial and seasonal springs are connected to the Senchal Lakes, which were built in 1910 and 1932. After purification at a filtration plant in Jorebungalow, the water is piped from Senchal to the town. The water distribution system within Darjeeling comprises a network of pipes totaling 83 kilometers, in addition to the 35 kilometers of pipelines that transport water from Senchal to Darjeeling. During the pre-monsoon months when the water level in the Senchal lakes decreases, water is supplemented by electrically pumping it from another reservoir near KhongKhola.

The system was initially designed to cater to a maximum of 20,000 individuals. However, between 1911 and 2011, the population of the municipality increased by six times, excluding the significant number of temporary residents such as students, migrant workers, and tourists. The growing demand has resulted in a worsening shortage of water supply. Consequently, numerous residents are forced to buy water from private sellers who deliver it in tankers or manually transported carts. In some cases, people collect water from local streams or springs. Bigger private companies also provide water to households but at significantly higher prices.

Area Irrigated bv different *Source*s in 1 he district of Darjeeling

Year in	AIHH iftnalMl hv									
	■ -■■I ..L-LL	Tank	unrw	MDT1AI	inrw	STW	RLI	nnw	CHhpJs.	Tiilal
			Ui	IW	(Rl	in	(RI	IQl	itm	Mil
2008-09	5.94	*	•	•	•	062	014	018	040	7.28
2009-10	410	*	•	*	•	039	016	0.18		483
2010-11	2.83	*	•	•	*	0.39	0.16	0.18		356
2011-12	392	*	*	•	*	012	0.30		048	482
	3,92	•		0.08'		018	472		164	10,54

Note HDTW Duro Titwwrt!
MQTW Mdcfc cawoh Dm Tubwwi
LDTW Lo* Gfldontv Dceo Tufcwvi
STW SMow Tutwei RLI- River Lfl Imottoc
OtM OwDwVW 'DTW=HOTW»MDTW*LDriV

Fig. 4.3.5.1: Area Irrigated by different *Source*s in Darjeeling

4.4 Profile of the Block

4.4.1 Geography of the Block

The Darjeeling-Pulbazar C.D. Block lies between 27°05′25″N 88°13′18″E. The northernmost portion of the district is covered by this area, which has the greatest elevation. This area includes the Singalila Ridge, the tallest mountain range in the district and the state. Sandakphu 11,929 ft (3,636 m) and Phalut 11,800 ft. (3,600 m) are the highest points in West Bengal.[2] The West Sikkim district and South Sikkim district of Sikkim are located on the north, RangliRangliot CD Block is located on the east, JorebunglowSukhiapokhri CD Block is located on the south, and Taplejung District of Province No. 1 of Nepal is located on the west.

The Darjeeling Pulbazar CD block has an area of 416.00 sq. km. It contains 1 panchayatsamity, 23 gram panchayats, 241 gram sansads (village councils), 46 mouzas, 43 inhabited villages, and 4 census towns. This block is serviced by the Pulbazar and Lodhoma police stations. The headquarters of this CD block is at Bijanbari.

The JorebunglowSukhiapokhri CD block is bordered on the north by the Darjeeling Pulbazar CD block, on the east by the CD blocks of RangliRangliot and Kurseong, on the south by the CD block of Mirik, and on the west by the CD blocks of Panchthar and Ilam in Nepal's Province No. 1.

The JorebunglowSukhiapokhri CD block has an area of 222.12 sq. km in size. There are 42 inhabited villages, 1 PanchayatSamity, 16 gram panchayats, 198 gram sansads, 47 mouzas, and 4 census towns. This block is served by the police stations at Jorebunglow and Sukhiapokhri. Sukhiapokhri is home to the headquarters of this CD block.

4.4.2 Demographics of the Block

4.4.2.1 Population

Based on the 2011 Census of India, the Darjeeling Pulbazar CD block recorded a total population of 126,935 individuals. Out of this, 105,150 resided in rural areas, while 21,785 lived in urban areas. The population consisted of 63,828 males and 63,107 females, each accounting for 50% of the total. The age group of 0 to 6 years had a population of 11,696 individuals.

The following census towns were identified within the Darjeeling Pulbazar CD block in 2011, along with their respective populations: Badamtam Tea Garden (6,102), Ging Tea Garden (4,089), Chongtong Tea Garden (5,802), and Singtam Tea Garden (5,792).Additionally, the Darjeeling Pulbazar CD block included several large villages with a population exceeding 4,000 people, as per the 2011 census: Rimbick (6,980), Kaijalia (4,150), Goke (9,100), Bijanbari (5,338), Tukvar Tea Garden (4,791), and Lebong& Mineral Spring Tea Garden (6,236).

Other villages present in the Darjeeling Pulbazar CD block, along with their respective populations from the 2011 census, are as follows: Pattabong Tea

Garden (2,347), Lodhoma (703), Lebong Tea Garden (2,276), Relling (3,568), Jhepi (1,576), Majua (1,164), Rungneet Tea Garden (1,252), Singla Tea Garden (3,400), Phubsering Tea Garden (3,208), Soom Tea Garden (3,578), Bloomfield Tea Garden (2,901), and Rishihat Tea Garden (1,651).

Based on the 2011 Census of India, the JorebunglowSukhiapokhri CD block had a total population of 113,516 individuals. Out of this population, 86,637 resided in rural areas, while 26,879 lived in urban areas. The gender distribution was almost equal, with 56,232 (50%) males and 57,284 (50%) females. The number of individuals in the age group of 0 to 6 years was 9,313.

The census towns within the JorebunglowSukhiapokhri CD block, along with their respective populations in the 2011 census, were as follows: Sukhiapokhri (4,450), SonadaKhasmahal (11,635), RongmookCeder Tea Garden (5,150), and Mangarjung Tea Garden (Nagri) (5,644).Large villages in the JorebunglowSukhiapokhri CD block, with populations exceeding 4,000, were Dooteria Tea Garden (5,113) and Ringtong Tea Garden (Margaret's Hope) (6,025) according to the 2011 census.

Additionally, other villages within the JorebunglowSukhiapokhri CD block included Kalej Valley Tea Garden (2,640), MoondaKotee Tea Garden (3,972), PokhribongKhasmahal (3,051), GopalDhara Tea Garden (2,084), Rangbul (3,916), Jorebunglow (2,854), Mariabong Tea Garden (2,404), Pushinbong Tea Garden (2,879), Poobong Tea Garden (2,170), Nahori Tea Garden (Balason) (2,402), Sagmaru Tea Garden (Sungma) (2,229), Tarzun Tea Garden (2,232), Molatey Tea Garden (Aveneroge) (2,285), Nagri Farm Tea Garden (3,655), Dhajea Tea Garden (1,394), and Tumsong Tea Garden (1,700), with their corresponding population figures from the 2011 census.

4.4.2.2 Literacy

According to the 2011 census the total number of literate persons in the Darjeeling Pulbazar CD block was 93,091 (80.78% of the population over 6 years) out of which males numbered 50,630 (87.61% of the male population over 6 years) and females numbered 42,461 (73.91% of the female population over 6 years). The gender disparity (the difference between female and male literacy rates) was 13.69%.

Similarly the total number of literate persons in the JorebunglowSukhiapokhri CD block was 86,004 (82.54% of the population over 6 years) out of which males numbered 46, 323 (89.98% of the male population over 6 years) and females numbered 39,681 (75.26% of the female population over 6 years). The gender disparity (the difference between female and male literacy rates) was 14.72%.

4.4.2.3 Language and Religion

Nepali is the predominant language, with 96.71% of the population. Various Bhutia and Tibetan dialects are spoken by 1.17% of the population.

In the 2011 census, Hindus numbered 83,166 and formed 65.52% of the population in the Darjeeling Pulbazar CD block. Buddhists numbered 24,712 and formed 19.47% of the population. Christians numbered 1,1897 and formed 9.37% of the population. Muslims numbered 569 and formed 0.45% of the population. Others numbered 6,541 and formed 5.19% of the population.

At the time of the 2011 census, 95.53% of the population spoke Nepali, 1.12% Hindi and 1.01% Bhojpuri as their first language in the Jorebunglow - Sukhiapokhri C.D. Block.

In the 2011 census, Hindus numbered 77,014 and formed 67.84% of the population in the JorebunglowSukhiapokhri CD block. Buddhists numbered 29,680 and formed 26.15% of the population. Christians numbered 5,084 and formed 4.48% of the population. Muslims numbered 699 and formed 0.62% of the population. Others numbered 1,039 and formed 0.92% of the population.

4.4.2.4 Economy

The economy of Darjeeling relies heavily on three main sectors: tourism, the tea industry, and timber. Local residents who own hotels and restaurants provide employment opportunities for the community, while others earn a living by selling local handicrafts, souvenirs, and working as tourist drivers. Tourism plays a vital role in the region's economy, generating income and employment for the local population. Annually, tourists spend around Rs. 70 crore in the area.

In addition to tourism, the tea industry is the largest industrial activity and employer in the hilly region. There are currently 87 registered tea gardens covering an area of 17,500 hectares, with 11 of them being organic gardens and 2 facing challenges. The annual tea production ranges from 10 to 11 million kg. The industry provides direct and indirect employment to approximately 50 percent of the local population, with 60 percent of the directly employed workers being women. Employment in the tea industry follows a family-based system, often employing workers from the third or fourth generation. The Darjeeling tea industry sustains over 55 thousand permanent employees throughout the year, and an additional 25 thousand workers are engaged during the plucking season, which extends from March to November.

4.4.2.5 Tea Economy

The tea industry is not only crucial for the people living in the hills of Darjeeling but also serves as the backbone of the local economy (Lama, 1986). A significant number of tea estates are located in the Darjeeling and Kurseong subdivisions. In contrast, the Kalimpong subdivision has fewer tea plantations due to its focus on forest reserves and agricultural activities carried out by farmers. By 1905, the tea industry had become the primary industry in Darjeeling, with approximately one-third of the entire population depending on the tea estates (Sarkar and Lama, 1986).

The tea industry holds considerable importance in the national economy of India. India maintains its position as the world's largest producer, consumer, and exporter of tea (Roy, 1968; Misra, 1986). India's prominent role in tea production, consumption, and export contributes significantly to various aspects such as gross national product (GNP), export earnings, revenue re*Source*s, job creation, regional development, human potential growth, ecological balance, and the immeasurable value of a popular beverage. The dual nature of tea, as both an agricultural and industrial enterprise, has given rise to two separate attitudes, each operating independently of the other (Sarkar, 1984).

4.4.2.6 Tourism

Tourism in mountainous regions is primarily promoted for economic purposes. It offers the potential for generating financial re*Source*s in remote mountain areas that have limited economic opportunities. Additionally, tourism serves as a *Source* of local employment, helping to mitigate the issue of depopulation in marginalized areas. Furthermore, it contributes to addressing regional imbalances by redirecting economic activities and opportunities to these regions (Singh *et al*., 1992).

Tourism plays a significant role in the local economy of this mountainous area by generating income and providing employment opportunities for the local population. Tourists annually spend approximately seventy crores in this region. The impact of tourism extends beyond increased income and demand for local products. It also leads to a multiplier effect, which refers to the way tourist expenditures filter through the economy and stimulate additional economic activities. The multiplier effect is based on the interdependency among different sectors of the economy, meaning that any change in output, income, employment, government revenue, or foreign exchange flows in the host economy will be greater than the initial change in tourist spending. The multiplier is expressed as a ratio of the change in one of the aforementioned variables to the change in tourist spending that caused it. Moreover, tourism and related industries, being labour-intensive, have a significant impact on employment levels in the sector.

Income and employment generation are the primary positive impacts of tourism (Mashqura and Lepcha, 2004). Previously reliant on subsistence farming, local workers have now become entrepreneurs and businessmen, hiring guides and renting supplies to tourists visiting the mountains. People of all ages can contribute to the local economy by crafting souvenirs like baskets and clothes, which are eagerly purchased by tourists. Furthermore, tourism can bring about improvements in rural infrastructure in India. Besides the commerce generated through the construction of roads, various countryside areas have seen developments such as sewage systems, telephones, electricity, and other infrastructure related to tourism.

4.5 Location of the Study Area

The Darjeeling Sadar Subdivision, which encompasses the southern half of the Darjeeling district in West Bengal, includes the rural parts of Darjeeling- Pulbazar and Jorebunglow- SukhiaPokhri CD Blocks (Maps 1.1, 1.2), two of the nine blocks and four subdivisions that make up the Darjeeling District. The coordinates of Darjeeling- Pulbazar CD Block lies between 27°05′25″N 88°13′18″E. The northernmost portion of the district is covered by this area, which has the greatest elevation. This area includes the Singalila Ridge, the tallest mountain range in the district and the state. Sandakphu 11,929 ft (3,636 m) and Phalut 11,800 ft. (3,600 m) are the highest points in West Bengal. The West Sikkim district and South Sikkim district of Sikkim are located on the north, RangliRangliot CD Block is located on the east, JorebunglowSukhiapokhri CD Block is located on the south, and Taplejung District of Province No. 1 of Nepal is located on the west.JorebunglowSukhiapokhri is a Community development block (CD) that lies between 27°00′N 88°10′E.The Jorebunglow-Sukhiapokhri CD block is bordered on the north by the Darjeeling -Pulbazar CD block, on the east by the CD blocks of Rangli- Rangliot and Kurseong, on the south by the CD block of Mirik, and on the west by the CD blocks of Panchthar and Ilam in Nepal›s Province No. 1.

4.6 Administrative Setup of the Study Area

The Darjeeling Pulbazar CD block has an area of 416.00 sq. km. It contains 1 panchayatsamity, 23 gram panchayats, 241 gram sansads (village councils), 46 mouzas, 43 inhabited villages, and 4 census towns. This block is serviced by the Pulbazar and Lodhoma police stations. The headquarters of this CD block is at Bijanbari.The JorebunglowSukhiapokhri CD block has an area of 222.12 sq. km in size. There are 42 inhabited villages, 1 PanchayatSamity, 16 gram panchayats, 198 gram sansads, 47 mouzas, and 4 census towns. This block is served by the police stations at Jorebunglow and Sukhiapokhri. Sukhiapokhri is home to the headquarters of this CD block.The research area covers 638.12 sq. km of geographical area. There are four Subdivisions in Darjeeling District, out of which Darjeeling Sadar Subdivision primarily consists of the following administrative units.

1. Darjeeling -Pulbazar CD Block
2. Darjeeling (Sadar) Municipality
3. Rangli- Rangliot CD Block
4. Jorebunglow -SukhiaPokhri CD Block

The study area which is composed of 39 Gram Panchayats(23 in Darjeeling-Pulbazar and 16 in Jorebunglow- SukhiaPokhri CD Blocks)[Table 4.6.3 and 4.6.4]. Darjeeling-Pulbazar CD Block has 47 villages(4 consensus towns) and Jorebunglow-SukhiaPokhri CD Block has 46 villages(4 consensus towns).

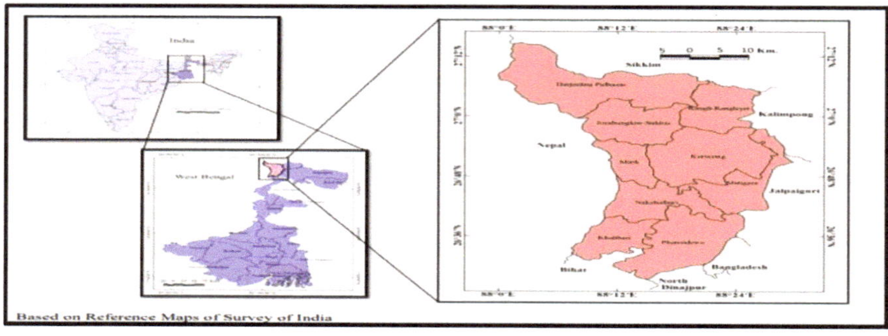

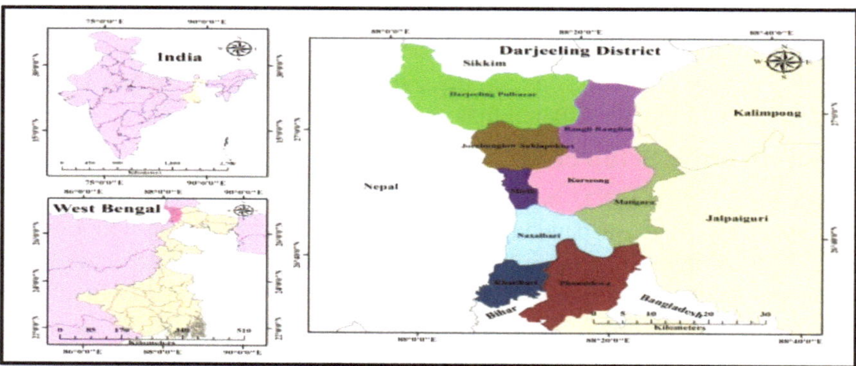

Fig. 4.5.1: Location Map of the Study Area

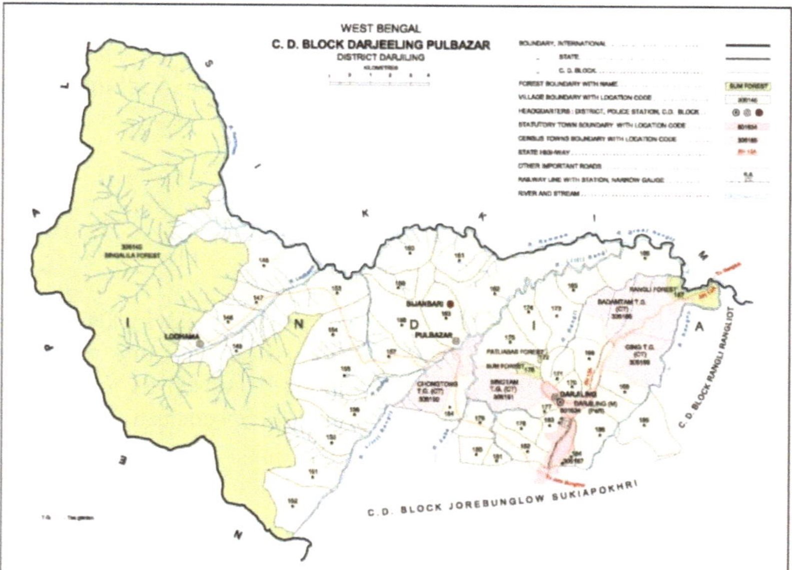

Fig. 4.5.2: Map of Darjeeling Pulbazar CD Block

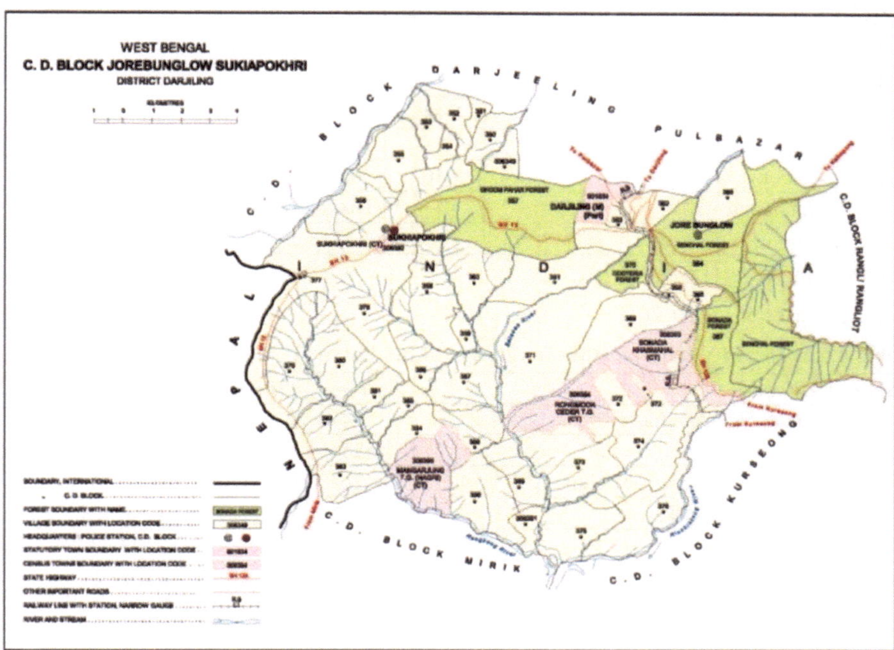

Fig. 4.5.3: Map of Jorebunglow Sukhiapokhri CD Block

Source: Administrative Atlas, West Bengal, 2011, Census of India

Table 4.6.1: List of Villages in Darjeeling-Pulbazar C.D Block

Sl no.	*Villages*	*Sl no.*	*Villages*
1	Badamtam Tea Garden *	25	Lebong& Mineral Spring Tea Garden
2	Ging Tea Garden*	26	Liza Hill Tea Garden
3	Chongtong Tea Garden*	27	Lodhoma
4	Singtam Tea Garden*	28	Majua
5	Alubari Busty	29	Murmidong
6	Alubari Tea Garden	30	Namla
7	Arya Tea Garden	31	Pandam Tea garden
8	Banockburn Tea Garden	32	Patliabas Forest
9	Barbatia khasmahal	33	Pattabong Tea Garden
10	Barnesbeg Tea Garden	34	Phubsering Tea Garden
11	Bijanbari	35	Rangli Forest
12	Bloomfield Tea Garden	36	Relling
13	Dangia	37	Rimbick
14	Darjiling	38	Rishihat Khasmahal
15	Goke	29	Rishihat Tea Garden
16	Happy Valley Tea Garden	40	Rungneet Tea Garden
17	Hatta	41	Samalbong

Sl no.	*Villages*	*Sl no.*	*Villages*
18	Jhepi	42	Singalila Forest
19	Kaijalia	43	Singbhumdera
20	Kankibong	44	Singla Tea Garden
21	Karmi	45	Soom Tea Garden
22	Kolbong	46	Soom Forest
23	Lamagaon	47	Tukvar Tea Garden
24	Lebong Tea Garden		

*Census Towns (2011)

Source: Census of India (2011)

Table 4.6.2: List of Villages in Jorebunglow- Sukhia Pokhri C.D Block

Sl no.	*Villages*	*Sl no.*	*Villages*
1	Sonada Khasmahal*	24	Parmaguri Khasmahal
2	Mangarjung Tea Garden*	25	Pokhribong Khasmahal
3	Rongmook Ceder Tea Garden*	26	Pubong Tea Garden
4	Sukhia Pokhri Town*	27	Pulung Dong Khasmahal
5	Achhalal Hatta	28	Pulung Dong Tea Garden
6	Chamu Tea Garden	29	Pussibong Tea Garden
7	Dhajea Khasmahal	30	RangBul
8	Dhajea Tea Garden	31	Rangaroong Tea Garden
9	Dooteria Tea Garden	32	Ringtong Tea Garden
10	Ghoom Pahar Forest	33	Ring Tong Tea Garden
11	Gopal Dhara Tea Garden	34	Sagmaru Tea Garden
12	Hill Card Road	35	Salu Tea Garden
13	JoreBunglow	36	Samrik Tea Garden
14	Kacel Tea Garden	37	Senchal Forest
15	Kalej Valley Tea Garden	38	Seyok Tea Garden
16	Lingia Tea Garden	39	Simana Basti
17	Mariabong Tea Garden	40	Sonada Forest
18	Mim Tea Garden	41	Soolbongs Tea Garden
19	Molatey Tea Garden	42	Tarzum Tea Garden
20	Moonda Kotee Tea Garden	43	Tumsong Khasmahal
21	Nagri Farm Tea Garden	44	Tumsong Tea Garden
22	Nahori Tea Garden	45	Dooteriah Tea Garden
23	Okas Tea Garden	46	Mim Nagrik Range

*Census Towns(2011)

Source: Census of India (2011)

Table 4.6.3: List of Gram Panchayats in Darjeeling-Pulbazar C.D. Block

Sl.no	*Gram Panchayats*	*Sl. no*	*Gram Panchayats*
1.	Badamtam	13.	Lodhoma I
2.	Bijanbari-Pulbazar	14.	Lodhoma II
3.	Chungtong	15.	Majua
4.	Dadaipani	16.	Nayanor
5.	Darjeeling I	17.	Rangit I
6.	Darjeeling II	18.	Rangit II
7.	Goke I	19.	Relling
8.	Goke II	20.	Rimbick
9.	Jhepi	21.	Singtam
10.	Kaijalia	22.	Soom
11.	Lebong Valley I	23.	Sirikhola-Daragaon
12.	Lebong Valley II		

Source: District Statistical Handbook, Darjeeling 2001-2011.

Table 4.6.4: List of Gram Panchayats in Jorebunglow-Sukhia Pokhri C.D. Block

Sl.no	*Gram Panchayats*	*Sl.no*	*Gram Panchayats*
1.	Dhotrey kalej Valley	9.	Pokhriabong I
2.	Ghoom Khasmahal	10.	Pokhriabong II
3.	Gorabari Margarets Hope	11.	Pokhriabong III
4.	Lingia Marybong	12.	Rangbhang Gopaldhura
5.	Lower Sonada II	13.	Rangbul
6.	Mundakothi	14.	Sukhia Simana
7.	Permaguri Tamsang	15.	Upper Sonada
8.	Pulung Dung	16.	Lower Sonada I

Source: District Statistical Handbook, Darjeeling, 2001-2011.

The structure of Panchayats in Darjeeling-Pulbazar C.D. Blocks is composed of one Panchayat Samiti, 23 Gram Panchayats and 241 Gram Sansads. Similarly, the structure of Panchayats in Jorebunglow-Sukhia Pokhri C.D. Block is composed by one Panchayat Samiti, 16 Gram Panchayats, 198 Gram Sansads (District Statistical Handbook, Darjeeling, 2011).

Fig. 4.5.1: Bird's eye view and Satellite view of Pulungdung village

Gram Panchayat: PulungdungKhasmahal

Block/ Subdivision: Jorebunglow- Sukhiapokhri

District: Darjeeling

State: West Bengal

Pincode: 734221

Area: 628.08 hectares

Population: 2573

Households: 533

Fig. 4.5.2: Bird's eye view and Satellite view of Tukver Tea Garden village

Gram Panchayat:Rangit I

Block/ Subdivision: Darjeeling Pulbazar

District:Darjeeling

State: West Bengal

Pincode:734124

Area: 634.15 hectares

Population: 4791

Households: 1034

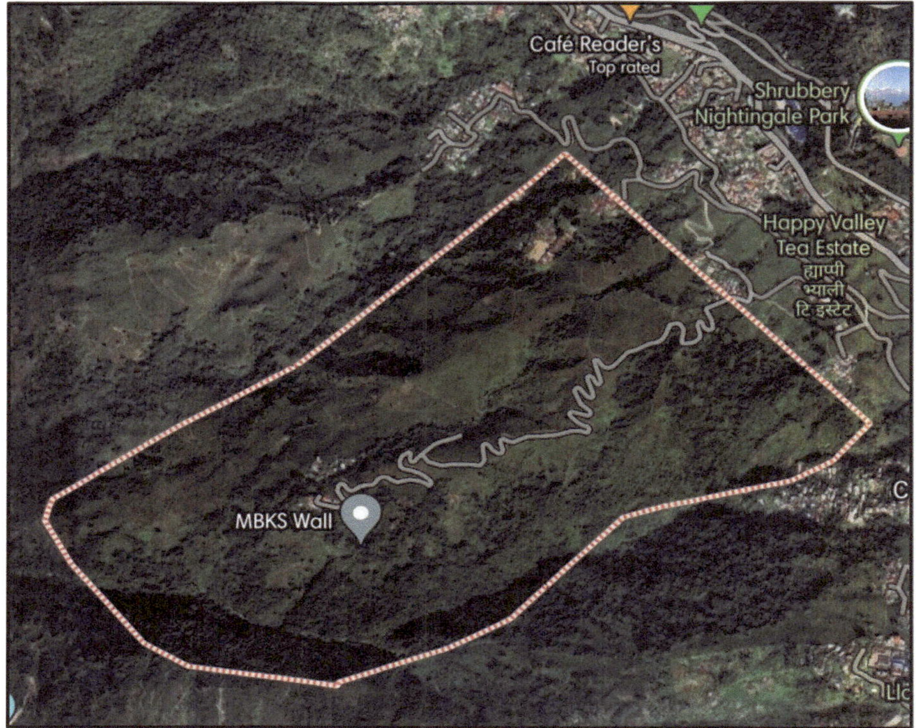

Fig. 4.5.3: Bird's eye view and Satellite view of Happy Valley Tea Garden village

Gram Panchayat: Darjeeling I

Block/ Subdivision:Darjeeling- Pulbazar

District: Darjeeling

State:West Bengal

Pincode:734101

Area: 177 hectares

Population: 1500

Households:303

Fig. 4.5.4: Bird's eye view and Satellite view of Pokhriabong village

Gram Panchayat:Pokhriabong I

Block/ Subdivision:Jorebunglow-Sukhiapokhri

District:Darjeeling

State:West Bengal

Pincode:734216

Area: 269.93 hectares

Population:3051

Households:610

5

Methods and Material

The methodology has been carefully considered in order to comprehend the idea, procedures, and approaches used in the design of the study, information gathering, data analysis, and finding interpretation for the identification of relevant truths and the development of hypotheses. The methodology and a process employed in the study are covered in this chapter.

Any endeavour that demands tangible results absolutely must have a well-defined process. The scientific methodology used to carry out this research study is described in this chapter and is offered under the following sub heading:

1. Locale of research.
2. Pilot Study.
3. Sampling design.
4. Item analysis and Experts rating for deviating scoring technique on dependent variables.
5. Preparation of Interview Schedule.
6. Pre-testing of Schedule.
7. Techniques of Data Collection.
8. Empirical Measurements of the Variables.
9. Statistical Tools for Analysis of Data.

5.1 Locale of the Research

The rural areas of Darjeeling-Pulbazar CD block, Jorebunglow-SukhiaPokhri CD block in Darjeeling district in West Bengal was selected for the study. The area had been selected for the study because of the availability of appropriate respondents, people as well as the local language.

5.2 Pilot Study

Prior to beginning the actual research, a pilot study was carried out to better understand the regions, its inhabitants, institutions, and programmes activities associated to the migration process in the research area. During the pilot study period, the respondents' situational and historical data were gathered.

5.3 Sampling Design

Techniques for multiple, simple, and purposive random sampling were used in the study. Purposive sampling techniques were used for the selection of the state, district, block, and villages since the location were perfect for the problem.

Table 5.3.1: Sampling Design and Sampling Techniques

Sl.no.	Level	Approach	Remarks
1.	State- West Bengal	Purposive	To identify the different factors or variables responsible for migration and relationship between them in the Darjeeling region.
2.	District- Darjeeling	Purposive	
3.	Block- Darjeeling Pulbazar; Jorebunglow- Sukhiapokhri.	Purposive	The area was ideal with respect to the problem, convenient for the researcher and having the infrastructural facilities.
4.	Villages- Pulungdung Busty, Pokhribong, Tukver Tea Garden, Happy Valley Tea Garden.	Purposive	Ease of accessibility, availability of respondents, substantial rural population as well as local language.
5.	20 respondents from each villages.	Random and non random	Collection of data from 80 respondents

5.4 Item analysis and Experts rating for deviating scoring technique on dependent variables

Since the topic of the research has been so new and unique, the current researcher under the guidance of the supervisor has generated a score of marker questions to be rated on ten point scale by the selected experts comprising of eminent scientist in the relevant areas and extension practitioner widely exposed to the domain. The steps were as follows.

a. Random selection of item.
b. Screening and scrutinizing of the items based on applicability and relevancy.
c. Finalization of marker item/ statements to be rated on ten point scale.
d. Selection of experts and seeking their consents.
e. Sending the item to the selected experts and collecting their feedback (80 per cent of the feedback from the experts were available).
f. Respected item values were summated.
g. 50 percent of the items were retained and rest were discarded.
h. Through before and after test on 20 respondents, the r value was calculated and it was found significant.
i. The final marker statement has thus been finally selected and been applied for data collection.

5.5 Preparation of Interview Schedule

Based on the results of the pilot study, a preliminary interview schedule was created with the aid of literature, with the chairman of the advisory committee's guidance, and after consultation with the advisory committee's members.

5.6 Pre-testing of Schedule

Prior to commencing the final data collection, entire schedule was presented to the study area's programme provider respondents for elimination, substitutions, additions and modifications to the schedule.

5.7 Techniques of Data Collection

This was personally interviewed and collected during Puja vacation of 2022(especially in the time of Dasai, Lakhsmi Puja and Diwali festival where the maximum migrants return to their homes) and summer vacation of 2023. The questions were asked in the local language so that the respondent can easily understand the queries.

5.8 Empirical Measurements of the Variables

After reviewing various literatures related to the field of the study and consultation with respected Chairman of Advisory Committee and other experts, a list of variables was prepared, on the basis of which a schedule was formed.

5.8.1 Dependent Variables

Considering the objective of the study, (a) perceived benefits of migration (y_1) (according to the respondent's, five important factors determine the extent of benefit from the migration process, the five factors are social esteem, economic satisfaction, job satisfaction, better access to education and healthcare, improved quality life), (b) Push factors (y_2) are generally the aspects or conditions that motivates one to leave their own place, region or organization, (c) pull factors (y_3) are the aspects or conditions that attracts the migrants to move to new locations. According to the respondents perception there are some push and pull factors which play crucial role in the migration process. For determining the real reason of migration the perceived push and pull factors are identified with the aid of respondents and accordingly the perceived important factors contributing to the migration process the respondents give scoring to each of the factor out of ten.

Table 5.8.1.1: List of Dependent Variables

Sl.No.	Dependent Variables		Notations
1.	Perceived Benefits.		Y_1
	Factors	*Scoring out of 10*	
	Social esteem		
	Economic satisfaction		
	Job satisfaction		
	Better access to Education & Healthcare		
	Improved Quality life		
2.	Push Factors.		Y_2
	Factors	*Scoring out of 10*	
	Poverty		
	Low net return from agriculture		
	Insecurity and risk		
	Low social status		
	Improved connectivity		
	Irregularity of income		
	Price rise		
	Lack of local market		
	Influence of family		
	Lack of versatility of working sector		
3.	Pull Factors.		Y_3
	Factors	Scoring out of 10	
	High wage		
	Desire for better lifestyle		
	Improved communication network		
	High social esteem		
	Regularity of income		
	Influence by villagers		
	Job driven migration		
	Influence by friends		
	High return on per hour labour investment		
	Better job opportunity		
	Skill development		

5.8.2 Independent Variable

Age(x_1), Marital Status(x_2), Age at time of migration(x_3), Education (Respondent) (x_4), Family size(x_5), Land holding (x_6), Social participation(x_7), Pattern of Migration(x_8), Family income (x_9), No. of *Source* of information for migration(x_{10}), Social support network at destination (x_{11}), No. of income *Source* (x_{12}), Frequency of visiting home (x_{13}), Duration of stay at home (x_{14}), Distance of migration (x_{15}), Duration of migration (x_{16}).

Table 5.8.2.1: List of Independent Variables

Sl.No.	Independent Variables	Empirical Measurement
1	Age (x_1)	Chronological age
2	Marital Status (x_2)	Years
3	Age at the time of migration (x_3)	Chronological age
4	Education (x_4)	No. of years of exposure to formal education
5	Family Size (x_5)	Count
6	Land Holding (x_6)	Acre
7	Social Participation (x_7)	Involvement
8	Pattern of Migration (x_8)	Count
9	Family income (x_9)	Rupees/month
10	No. of information *Source* for migration (x_{10})	Count
11	Social support network at destination (x_{11})	Count
12	No. of income *Sources* (x_{12})	Count
13	Frequency of visiting home (x_{13})	Count
14	Duration of stay at home (x_{14})	No. of days
15	Distance of Migration (x_{15})	Kilometres
16	Duration of Migration (x_{16})	Years

Table 5.9: Statistical Tools used for Analysis of Data

Sl.No.	Approaches	Statistical Tools
1.	Descriptive visualization	Mean, Maximum, Minimum, Range Standard, Standard Deviation, Coefficient of Variation
2.	Influence Analysis	Coefficient of Correlation
3.	Direction Analysis	Path Analysis
4.	Functionality Analysis	Multiple and Stepwise Regression
5.	Conjoint	Canonical Covariate Analysis
6.	Dimension reduction	Factor analysis

5.8.3 Mean

Mean or $(\overline{X})=\sum\frac{xi}{n}$

Where, $\text{Standarddeviation}\ (\sigma)=\sqrt{\Sigma(x_i-\overline{x})^2/\text{n}}$ - the symbol we use for the mean (pronounced as X bar)

Σ = Symbol for summation

Xi = Value of the ith item X, I = 1,2,,n

N= total number of items

The concept of central tendency, which includes statistical averages, provides information about the central point around which data tends to cluster. This measure is considered the most representative value for the entire dataset. The measure of central tendency is also referred to as a statistical average. The most commonly used averages are mean, median, and mode. Mean, also known as the arithmetic average, is the most frequently employed measure of central tendency. It can be calculated by dividing the sum of all the given values in a series by the total number of items in that series.

The mean is a straightforward and widely utilized measure of central tendency. Its primary purpose is to summarize the key characteristics of a dataset and facilitate comparisons between data. It lends itself well to algebraic manipulation and is employed in various statistical calculations. As a measure of central tendency, it tends to be relatively stable. However, it does have some limitations. For example, it can be heavily influenced by extreme values, it may not align with the actual value of an individual item in a series, and it can create strong impressions, especially when the individual values are not provided alongside the average. Nevertheless, the mean is considered superior to other averages, particularly in economic and social studies where direct quantitative measurements are feasible.

5.8.4 Maximum and Minimum

In statistics, the sample maximum and sample minimum are also known as the largest observation and smallest observation, respectively. They represent the greatest and least values found within a sample. These values serve as fundamental summary statistics and are commonly used in descriptive statistics, including the five-number summary, Bowley's seven-figure summary, and associated box plots.

The sample maximum and sample minimum can be considered the first and last order statistics, denoted as X (1) and X(n), respectively, when the sample size is n.

When a sample contains outliers, it is inevitable that the sample maximum or sample minimum, or both, will include these outliers, depending on whether they are exceptionally high or low. However, it is important to note that the sample maximum and minimum do not necessarily represent outliers themselves unless they deviate significantly from the other observations.

5.8.5 Standard Deviation

The standard deviation is the most commonly employed measure to quantify the dispersion or variability within a dataset. It is typically represented by the symbol "σ" (pronounced as sigma). The standard deviation is calculated as the square root of the average of the squared deviations of individual values from the arithmetic average of the series. It is worked out as:

$$\text{Standard deviation } (\sigma) = \sqrt{\Sigma(x_i - \overline{x})^2 / \text{n}}$$

5.8.6 Variance

In probability theory and statistics, variance represents the expected value of the squared deviation of a random variable from its mean. It provides a measure of how spread out a set of random numbers is from their average value. Variance plays a central role in statistics and is utilized in various statistical concepts and techniques, including descriptive statistics, statistical inference, hypothesis testing, goodness of fit, and Monte Carlo sampling. In scientific fields where data analysis is common, variance serves as an important tool. It is calculated as the square of the standard deviation, which is the second central moment of a distribution and also the covariance of the random variable with itself. Variance is often denoted by symbols such as σ^2, s^2, or Var (X).

5.8.7 Coefficient of Variation

The coefficient of standard deviation is obtained by dividing the standard deviation by the arithmetic average of the series. It serves as a relative measure and is often used to compare with similar measures of other series. Multiplying this coefficient of standard deviation by 100 gives us the coefficient of variation. Additionally, the square of the standard deviation, known as variance, is frequently used in the context of analysing variance.

The standard deviation, along with related measures like variance and coefficient of variation, is primarily used in research studies and is considered a highly satisfactory measure of dispersion in a series. It can be mathematically manipulated as it takes into account the algebraic signs, unlike the mean deviation. Moreover, it is less influenced by sampling fluctuations. These advantages have contributed to the popularity of the standard deviation and its coefficient as measures of the spread or dispersion of a series. They are widely utilized in the estimation and testing of hypotheses.

5.8.8 Coefficient of Correlation

Up until now, we have focused on statistical measures applicable to univariate populations, which involve the analysis of a single variable. However, when dealing with bivariate or multivariate populations, we often aim to understand the relationship between two or more variables in the dataset.

Karl Pearson's coefficient of correlation, also known as simple correlation, is the most widely used approach to quantifying the degree of association between two variables. This coefficient operates under the following assumptions:

1. There is a linear relationship between the two variables.
2. The two variables are causally linked, with one variable acting as the independent variable and the other as the dependent variable.
3. Numerous independent causes are at play in both variables, resulting in a normal distribution.

The coefficient of correlation serves as a tool for assessing the strength and direction of the relationship between two variables, provided that the aforementioned assumptions hold.

Karl Pearson's coefficient of correlation

$r = \Sigma(X_i - \bar{X})(Y_i - \bar{Y}) / n.\ \sigma x.\ \sigma y$

Where, Xi = ith value of X variable

$\bar{X}$ = mean of X

Yi = Ith value of Y variable

$\bar{Y}$ - mean of Y

n = number of pairs of observations of X and Y

σx = Standard deviation of X

σy = Standard deviation of Y

Karl Pearson's coefficient of correlation, also referred to as the product moment correlation coefficient, is a measure used to quantify the relationship between two variables. The coefficient 'r' ranges between -1 and +1. A positive value of 'r' indicates a positive correlation, where both variables change in the same direction. Conversely, a negative value of 'r' indicates a negative correlation, with changes in the variables occurring in opposite directions. A value of 0 for 'r' indicates no association between the variables.

When 'r' approaches +1 or -1, it signifies a strong correlation between the variables. A value of +1 represents a perfect positive correlation, indicating that variations in the independent variable (X) account for 100% of the variations in the dependent variable (Y). In other words, for each unit change in the independent variable, there is a consistent change in the dependent variable in the same direction. Conversely, a value of -1 represents a perfect negative correlation, meaning that the changes in the variables occur in opposite directions.

A value of 'r' closer to +1 or -1 indicates a higher degree of correlation between the two variables.

5.8.9 Regression

The correlation coefficient only indicates the presence of an association between variables and does not provide information about causal relationships. Knowing that two variables, x and y, are correlated does not allow us to determine if the variation in x causes the variation in y, vice versa, or if the association is due to a mutual dependence or common causes affecting both variables. Additionally, a high correlation coefficient does not necessarily imply an underlying relationship between the variables. It is possible to have a strong correlation without a meaningful connection between the variables.

To understand the underlying relationship between y and x in a bivariate population, a mathematical equation called the regression equation is used. This equation represents the regression of variable y on variable x and captures the relationship between the two variables in a more comprehensive manner.

If y is the dependent variable and x is the independent variable, then the linear regression equation can be written as

$y = a+bx$

The values of a and b can be obtained by the method of least squares which consists of minimizing the expression.

Σ(y1-1-bxl)2 with respect to a and b

The values of a and b

$a = y\text{-}bx$

$$b = \frac{\Sigma xy - (\Sigma yi)}{\Sigma xi^2 \frac{(\Sigma xi)^2}{n}}$$

The regression line can now be written as

Y = y-bx + bx + bxory-y = b (x-x)

Where b is the regression coefficient.

5.8.10 Multiple Regression Analysis

When dealing with multiple independent variables, the analysis pertaining to the relationship between them is referred to as multiple correlations. The equation that describes this relationship is called the multiple regression equation. In this explanation, we will focus on the scenario where there are two independent variables and one dependent variable. It's worth noting that computer programs are available to handle a larger number of variables, making the analysis more convenient. In this situation the results are interpreted as shown below:

Multiple regression equation assumes the form

$\hat{Y} = a + b_1X_1 + b_2X_2$

Where X_1 and X_2 are two independent variables and Y being the dependent variable, and the constants a, b_1 and b_2 can be solved by solving the following three normal equations:

$\Sigma Yi = na + b_1\, \Sigma\, X_1i + b_2\Sigma\, X_2i$

$\Sigma\, X_1i\, Yi = a\, \Sigma\, X_1i + b1\Sigma\, X\, 2\, 1i + b_2\Sigma\, X_1i\, X_2i$

$\Sigma\, X_2i\, Y_2\, i = a\, \Sigma\, X_2i + b_1\Sigma\, X_1i\, X_2i + b_2\Sigma\, X_2i^2$

The number of normal equations utilized in multiple regression analysis depends on the number of independent variables. For instance, if there are two independent variables, three equations are used. If there are three independent variables, four equations are used, and so on.

In multiple regression analysis, the reliability of regression coefficients (such as b1, b2) diminishes as the correlation between the independent variables (such as X1, X2) increases. This presents a challenge known as multicollinearity when there is a high degree of correlation between independent variables. In such cases, it is recommended to use only one set of independent variables for estimation purposes. The addition of a second variable (e.g., X2) that is correlated with the first variable (e.g., X1) distorts the values of the regression coefficients. However, predictions for the dependent variable can still be made even in the presence of multicollinearity. Care should be taken when selecting independent variables to estimate a dependent variable to minimize the impact of multicollinearity.

In multiple regression analysis with more than one independent variable, a distinction can be made between the combined effect of the independent variables and their individual effects when considered separately. The combined effect is measured by the coefficient of multiple correlations.

In multiple regression, a linear combination of explanatory variables is formed to maximize its correlation with a criterion variable. The primary objective of this technique is to predict the variability of the dependent variable based on its covariance with all the independent variables. By employing multiple regression analysis models, one can predict the level of the dependent phenomenon by considering the levels of the independent variables.

5.8.11 Stepwise Multiple Regression

Stepwise regression is a variation of multiple regression that helps in selecting the most predictive independent variables while keeping the number of variables to a minimum. It allows for the step-by-step application of the least square method to solve one or more multiple linear regression problems. In each step of the analysis, a variable is added or removed based on its impact on reducing the error sum of squares. According to Draper and Smith (1981), the stepwise multiple regression method involves inserting variables one by one until the regression equation achieves satisfactory results. The order of variable insertion is determined by evaluating the partial correlation coefficient, which serves as a measure of the importance of variables that are not yet included in the equation.

The stepwise regression program, as described by Burroughs Corporation (1975), begins by constructing a correlation matrix. It identifies the best predictor, which is the independent variable with the highest correlation to the criterion variable, and performs a regression analysis using this predictor. The program then identifies the second-best predictor, the independent variable with the second-highest correlation to the criterion, and conducts a regression analysis using the multiple correlations of these two predictors. This process continues, adding predictors and performing regression analyses based on the correlations.It is important to note that at each step, the group of predictors being used may not necessarily be the best group of that size. In other words, the specific combination of independent variables used at a given step does not guarantee the highest multiple correlations with the criterion variable compared to any other group of the same size.

Instead, this group consists of variables that have the highest individual correlation with the criterion variable. The significance of a variable being considered for inclusion in the regression equation is evaluated using the F-statistic. If the F-value is below a certain threshold (less than the specified 'include' value), the variable is not added to the regression equation. The 'include' statement sets the minimum F-statistic required for a variable to be included. The significance of variables already present in the regression equation may change when new variables are added. The significance of these variables is also assessed using the F-statistic. If the F-value is below a specified threshold (less than the 'delete' value), the variable is not included in the equation. The 'delete' value establishes the F-statistic threshold below which a variable is removed from the regression equation.

To control degeneracy, which occurs when a variable introduced into the equation becomes a linear combination of variables already in the equation, a 'tolerance' level is specified. The tolerance statement sets the maximum value that a pivoted element can reach while still allowing its associated variable to be included in the equation. If the pivoted element associated with a variable falls below the specified tolerance level (in this example, 0.001), the variable is not included in the regression equation.

1. **Forward selection:** The forward selection method starts with an empty model and evaluates the statistical significance of each variable as it is added one by one. Variables that demonstrate the highest level of statistical significance are retained, and the process continues until the optimal results are achieved.
2. **Backward elimination:** In contrast, backward elimination begins with a model that includes all independent variables and progressively removes one variable at a time. After each elimination, the removed variable's statistical significance is tested to determine if it should be excluded from the model.
3. **Bidirectional elimination:** Bidirectional elimination combines elements of both forward selection and backward elimination. It tests the inclusion and exclusion of variables simultaneously, evaluating which variables should be added or removed based on their statistical significance. This method aims to strike a balance between the two approaches to identify the most appropriate set of variables for the model.

5.8.12 Path Analysis

The purpose of conducting Path Analysis is to gain a comprehensive understanding of the direct and indirect impacts of independent variables on the dependent variable. Path Analysis also allows for the identification of variables that serve as significant pathways for indirect effects.

In the context of Path Analysis, Singh and Chaudhary (1977) defined the path coefficient as the ratio of the standard deviation of the effect caused by a specific factor to the overall standard deviation of the total effect. For instance, if Y represents the effect and x1 represents the cause, the path coefficient for the path from x1 to Y is calculated as $\sigma x1/\sigma y$.

5.8.13 Canonical Correlation Analysis

Canonical correlation analysis is a statistical technique used to identify and measure associations between two sets of variables. It is employed in situations where multiple regression analysis is appropriate but with the presence of multiple inter-correlated outcome variables. The objective of canonical correlation analysis is to determine a set of canonical variates, which are orthogonal linear combinations of variables within each set that best explain the variability within and between the sets.

The essence of canonical correlation analysis can be defined as finding two sets of basis vectors, one for x and another for y, that maximize the correlations between the projections of the variables onto these basis vectors. It allows for the examination of relationships between sets of multiple dependent and independent variables. While multiple regression analysis predicts a single dependent variable from a set of independent variables, canonical correlation analysis simultaneously predicts multiple dependent variables from multiple independent variables. It imposes fewer restrictions on the types of data it can handle compared to other techniques, which are more limited in their applicability. As a result, some researchers consider canonical correlation analysis as a last resort, to be used when other higher-level techniques have been exhausted. However, in situations involving multiple dependent and independent variables, canonical correlation analysis is the most suitable and powerful multivariate technique.

Over time, canonical correlation analysis has gained acceptance in various fields and serves as a valuable tool for multivariate analysis, particularly as interest has expanded to encompassing multiple dependent variables.

To provide a more detailed explanation of canonical correlation, let's consider an extended example based on multiple regression analysis. Imagine a survey conducted to explore the connections between family size and income as predictors of the number of credit cards held by a family. In this scenario, the objective is to examine the relationship between two independent variables (family size and income) and a single dependent variable (number of credit cards).

Suppose the researcher's interest expanded to encompass a broader concept of credit usage. Instead of solely considering the number of credit cards held by a family, the researcher decided to incorporate another measure: the family's average monthly dollar charges on all credit cards. This approach provided a more comprehensive perspective on credit card usage. For readers interested in the use of multiple indicators to represent a concept, discussions of Factor Analysis and Structural Equation Modeling are recommended.

In this new scenario, the objective is to predict two dependent measures simultaneously: the number of credit cards and the average dollar charges. However, multiple regression analysis is limited to handling only a single dependent variable. Multivariate analysis of variance could be an option, but only if all independent variables were non-metric, which is not the case in this problem. Hence, canonical correlation analysis becomes the only technique available to examine the relationship with multiple dependent variables. By employing canonical correlation analysis, the researcher creates a composite measure of credit usage that incorporates both dependent variables. This eliminates the need to compute separate regression equations for each dependent variable. The outcome of the analysis is twofold: the identification of canonical variates, which represent the optimal linear combinations of independent and dependent variables, and the determination of the canonical correlation coefficient (Rc), which quantifies the strength of the relationship between the two sets of variables.

Canonical correlation analysis is a comprehensive member of the multivariate statistical techniques family. It shares similarities with various methods. It resembles regression in quantifying the relationship strength between the two sets of variables. It is akin to factor analysis in creating composite variables, and it bears resemblance to discriminant analysis in determining independent dimensions (similar to discriminant functions) for each variable set, aiming to achieve the highest correlation between these dimensions.

Canonical correlation analysis aims to determine the optimal structure or dimensionality of each variable set to maximize the relationship between independent and dependent variable sets. It examines the association between composite variables created from sets of multiple dependent and independent variables. In this analysis, several independent canonical functions are developed to maximize the correlation between the linear composites, also known as canonical variates, which represent the sets of dependent and independent variables. Each canonical function is based on the correlation between two canonical variates—one for the dependent variables and one for the independent variables. A distinctive aspect of canonical correlation analysis is that the variates are derived in a way that maximizes their correlation. This analysis does not stop at establishing a single relationship between the variable sets. Instead, multiple canonical functions (pairs of canonical variates) can be derived, allowing for a deeper exploration of the relationships between the sets of variables.

5.8.14 Factor Analysis

Factor analysis is a widely used and powerful statistical technique employed in various fields such as psychology, social sciences, market research, and data analysis. Its main purpose is to uncover and comprehend the underlying factors or latent variables that contribute to the relationships among a set of observed variables.

When faced with a large number of variables, factor analysis can be utilized to condense the information into a smaller set of factors. These factors capture common patterns or dimensions that explain the variability in the original variables. By reducing the dimensionality of the data, factor analysis simplifies the interpretation and analysis process.

The fundamental assumption of factor analysis is that the observed variables are influenced by a smaller number of unobserved factors. While these factors cannot be directly measured, they are inferred from the correlations observed among the measured variables. The technique aims to estimate the factor loadings, which represent the strength of the relationship between each observed variable and each factor. A high factor loading indicates a strong relationship, suggesting that the observed variable is influenced by that particular factor. After conducting factor analysis, the researcher examines the factor loadings and interprets the meaning of each factor. These interpretations depend on the specific context of the study and the domain of the variables involved. Factors can be labelled and described based on the variables that exhibit high loadings on each factor.

Factor analysis offers numerous advantages. It helps in simplifying data complexity, identifying the underlying structure of variables, and providing a more concise representation of the data. Additionally, it aids in reducing dimensionality for subsequent analyses such as regression or clustering. Moreover, factor analysis enables the discovery of hidden patterns, provides insights into the relationships among variables, and contributes to theory development.

It is crucial to understand that factor analysis is a statistical technique and does not establish causal relationships. It primarily serves as an exploratory method to enhance comprehension of the data and generate hypotheses for further investigation. Accurate interpretation of the results necessitates careful consideration of the study's context and theoretical background.

Table 5.10: Software Tools used for Analysis of Data

Sl.no.	Analysis	Software
1.	Descriptive and Inferential Statistics	Statistical Packages for Social Sciences(SPSS)
2.	Mental Modelling- Fuzzy Logic Cognitive Mapping	Mental Modeler

5.9 Statistical Packages for Social Sciences (SPSS)

Statistical Packages for Social Sciences (SPSS) is a software suite developed by IBM that provides a range of tools and capabilities for statistical analysis, data management, and data visualization. It is widely used in social sciences research, including fields such as psychology, sociology, economics, and political science. SPSS allows researchers to analyse and interpret complex data sets, conduct various statistical tests, and generate reports and charts to communicate the results.

SPSS offers a user-friendly interface, making it accessible to both novice and advanced users. It supports a wide range of statistical techniques, including descriptive statistics, t-tests, analysis of variance (ANOVA), regression analysis, factor analysis, cluster analysis, and many others. It also provides tools for data manipulation, data cleaning, and data transformation. SPSS supports both structured data, such as spread sheets and databases, and unstructured data, such as text files. It allows for data importing from various *Source*s, including popular file formats like Excel, CSV, and SQL databases. SPSS also provides features for data coding, recoding variables, and handling missing data.

The software includes a comprehensive set of visualization tools to create charts, graphs, and tables to present data in a visually appealing and meaningful way. Users can customize the appearance of visualizations and export them for use in reports or presentations. In addition to its core functionalities, SPSS allows for customization and extension through its programming language called Syntax. Users can write and execute Syntax commands to automate repetitive tasks, create complex analyses, and generate customized outputs.

5.10 Mental Modeler (Fuzzy-Logic Cognitive Mapping)

Mental Modeler is a computer software tool developed by the late Jay W. Forrester and his colleagues at the Massachusetts Institute of Technology (MIT). It is designed to help individuals and organizations create, analyse, and simulate causal loop diagrams and other types of system dynamics models. System dynamics is a discipline that studies the behaviour of complex systems over time. It explores the relationships between different variables within a system and how they interact to produce certain outcomes. Causal loop diagrams are graphical representations that show the cause-and-effect relationships among various variables in a system.

Mental Modeler provides a user-friendly interface for creating and editing causal loop diagrams. Users can define variables, specify their interconnections, and assign mathematical relationships to describe their behaviour. The software also allows users to simulate the model and see how changes in the variables impact the overall system behaviour.

Fuzzy-Logic Cognitive Mapping (FCM) is a parameterized form of concept mapping where you can develop qualitative static models that are translated into semi-quantitative dynamic models. Bart Kosko originally developed FCM in 1986 as a way to structure expert knowledge using a soft systems programming approach that is "fuzzy", thought to be similar to the way that the human mind makes decisions.

The analytical mechanics of FCM are based on examining the structure and function of concept maps, using graph theory-based analyses of pairwise structural relationships between the concepts included in a model. These models can be used to examine perceptions of an environmental or social problem or to model a complex system where uncertainty is high and there is little empirical data available.

Mental Modeler allows one to build Fuzzy-logic Cognitive Maps easily and intuitively. Once models are built, increasing or decreasing the components included in the model allows you to examine different scenarios of change. Because of their flexibility, FCM have been used in a range of scientific disciplines, from political science to economics to ecology.

Based in Fuzzy-logic Cognitive Mapping (FCM), users can easily develop semi-quantitative models of environmental issues, social concerns or social-ecological systems in *Mental Modeler* by:

- Defining the important components of a system
- Defining the relationships between these components
- Running "what if" scenarios to determine how the system might react under a range of possible changes.

The main purpose of Mental Modeler is to enhance understanding and decision-making by visualizing and analysing complex systems. By creating and simulating system dynamics models, users can explore the consequences of different scenarios, identify feedback loops, and gain insights into the behaviour of the system under study. Mental Modeler has been used in various fields, including business, public policy, environmental management, and education. It is a powerful tool for exploring the dynamics of complex problems and supporting the development of effective strategies and interventions.

6

The Case Studies and Visualization

In this chapter, the study's findings are presented in a structured manner, accompanied by a discussion based on the analysis. The study aimed to present its findings in alignment with its objectives. Towards the end of this chapter, connections between the findings are established, explanations are provided, and an effort is made to uncover the underlying causes. The results and their corresponding discussion are organized according to the specific objectives of the study.

Table 6.1: Descriptive Statistics of Independent Variable With Respect to Minimum, Maximum, Range, Mean, Standard Deviation of Values, Variance, and Coefficient of Variance

Variable	Minimum	Maximum	Mean	Std. Deviation	Variance	CV (%)
Age (x_1)	25.000	46.000	32.938	4.785	22.895	14.527
Marital Status (x_2)	0.000	18.000	3.838	3.973	15.783	103.527
Age at the time of migration(x_3)	18.000	30.000	22.825	2.894	8.374	12.678
Education (x_4)	0.000	16.000	10.425	3.791	14.374	36.367
Family Size (x_5)	2.000	7.000	4.313	1.176	1.382	27.261
Land holding (x_6)	0.230	1.220	0.598	0.263	0.069	44.068
Social Participation (x_7)	1.000	4.000	2.075	0.952	0.906	45.864
Pattern of Migration (x_8)	1.000	6.000	2.125	1.316	1.731	61.914
Family income (x_9)	13000.000	63000.000	30716.950	11169.651	124761101.162	36.363
No.of information *Source* for migration(x_{10})	1.000	5.000	2.463	1.272	1.619	51.668
Social support network at destination(x_{11})	1.000	4.000	2.125	0.817	0.668	38.454
No. of income *Sources*(x_{12})	1.000	3.000	1.700	0.644	0.415	37.903
Frequency of visiting home (x_{13})	1.000	4.000	2.163	0.906	0.821	41.909
Duration of stay at home(x_{14})	21.000	75.000	39.038	11.434	130.745	29.291
Distance of Migration(x_{15})	111.000	5467.000	1402.263	1125.430	1266592.069	80.258
Duration of Migration(x_{16})	1.000	20.000	9.213	4.324	18.701	46.941
Perceived Benefits(y_1)	5.600	8.200	6.633	0.622	0.386	9.371
Push Factors (y_2)	6.000	8.200	6.938	0.580	0.336	8.359
Pull Factors (y_3)	5.360	7.720	6.613	0.577	0.333	8.730

Result: Table 6.1 presents the distribution of 16 independent variables (x_1-x_{16}) and 3 dependent variables ($y_{1-}y_3$) in terms of their Minimum, Maximum, Mean, Variance, Standard deviation, and Coefficient of variation.

According to the study the maximum value for the independent variable **Age at survey (x_1)** is 46.00, and minimum is 25.00. The study's overall distribution was determined to have a mean of 32.938 with a variance of 22.895 and standard deviation 4.785. The high level of consistency is indicated by the variable's coefficient of variation, which is 14.527%.

According to the study the maximum value for the independent variable **Marital status (x_2)** is 18.00 and minimum is 0.00. The study's overall distribution was determined to have a mean of 3.838 with a variance of 15.783 and standard deviation 3.973. The low level of consistency is indicated by the variable's coefficient of variation, which is 103.527%.

According to the study the maximum value for the independent variable **Age at migration (x_3)** is 30.00, and minimum is 18.00. The study's overall distribution was determined to have a mean of 22.825 with a variance of 8.374 and standard deviation 2.894. The high level of consistency is indicated by the variable's coefficient of variation, which is 12.678%.

According to the study the maximum value for the independent variable **Education status (x_4)** is 16.00, and minimum is 0.00. The study's overall distribution was determined to have a mean of 10.425 with a variance of 14.374 and standard deviation 3.791. The high level of consistency is indicated by the variable's coefficient of variation, which is 36.367%.

According to the study the maximum value for the independent variable **Family size (x_5)** is 7.00, and minimum is 2.00. The study's overall distribution was determined to have a mean of 30716.950 with a variance of 1.382 and standard deviation 1.176. The high level of consistency is indicated by the variable's coefficient of variation, which is 27.261%.

According to the study the maximum value for the independent variable **Land holding (x_6)** is 1.220, and minimum is 0.230. The study's overall distribution was determined to have a mean of 0.598 with a variance of 0.069 and standard deviation 0.263. The fairly consistent level of consistency is indicated by the variable's coefficient of variation, which is 44.068 %.

According to the study the maximum value for the independent variable **Social Participation (x_7)** is 1.220, and minimum is 0.230. The study's overall distribution was determined to have a mean of 0.598 with a variance of 0.069 and standard deviation 1.316. The fairly consistent level of consistency is indicated by the variable's coefficient of variation, which is 45.864 %.

According to the study the maximum value for the independent variable **Pattern of Migration (x_8)** is 6.000, and minimum is 1.000. The study's overall distribution was determined to have a mean of 2.125 with a variance of 1.731 and standard deviation 0.263. The low level of consistency is indicated by the variable's coefficient of variation, which is 61.914 %.

According to the study, the maximum value for the independent variable **Family income (x_9)** is 63000.000, and minimum is 13000.000. The study's overall distribution was determined to have a mean of 30716.950 with a variance of 124761101.162 and standard deviation 11169.651. The high level of consistency is indicated by the variable's Coefficient of variation, which is 36.363 %.

According to the study the maximum value for the independent variable **No. of information** *Source* **for migration (x_{10})** is 5.000, and minimum is 1.000. The study's overall distribution was determined to have a mean of 2.463 with a variance of 1.619 and standard deviation 1.272. The fairly consistent level of consistency is indicated by the variable's Coefficient of variation, which is 51.668 %.

According to the study, the maximum value for the independent variable **Social support network at destination (x_{11})**, is 4.000, and minimum is 1.000. The study's overall distribution was determined to have a mean of 2.125 with a variance of 0.668 and standard deviation 0.817. The high level of consistency is indicated by the variable's coefficient of variation, which is 38.454 %.

According to the study, the maximum value for the independent variable **No. of income** *Source***s (x_{12})** is 3.000, and minimum is 1.000. The study's overall distribution was determined to have a mean of 1.700 with a variance of 0.415 and a standard deviation of 0.644. The high level of consistency is indicated by the variable's coefficient of variation, which is 37.903 %.

According to the study, the maximum value for the independent variable **Frequency of visiting home (x_{13})** is 4.000, and minimum is 1.000.The study's overall distribution was determined to have a mean of 2.163 with a variance of 0.821 and a standard deviation of 0.906. The high level of consistency is indicated by the variable's coefficient of variation, which is 41.909 %.

According to the study, the maximum value for the independent variable **Duration of stay at home (x_{14})** is 75.000, and the minimum value is 21.000.The study's overall distribution was determined to have a mean of 39.038 with a variance of 130.745 and a standard deviation of 11.434. The high level of consistency is indicated by the variable's coefficient of variation, which is 29.291%.

According to the study, the maximum value for the independent variable **Distance of migration (x_{15})** is 5467.000, and the minimum value is 111.000.The study's overall distribution was determined to have a mean of 1402.263 with a variance of 1266592.069 and a standard deviation of 1125.430. The less consistent level of consistency is indicated by the variable's coefficient of variation, which is 80.258 %.

According to the study, the maximum value for the independent variable **Duration of migration (x_{15})** is 20.000, and the minimum value is 1.000.The study's overall distribution was determined to have a mean of 9.213 with a variance of 18.701 and a standard deviation of 4.324. The fairly consistent level of consistency is indicated by the variable's coefficient of variation, which is 46.941 %.

According to the study, the maximum value for the Dependent variable **Perceived benefits (y_1)** is 8.200, and the minimum value is 5.600.The study's overall distribution was determined to have a mean of 6.633 with a variance of 0.386 and a standard deviation of 0.622. The high level of consistency is indicated by the variable's coefficient of variation, which is 9.371 %.

According to the study, the maximum value for the Dependent variable **Push factors (y_2)** is 8.200, and the minimum value is 6.000.The study's overall distribution was determined to have a mean of 6.938 with a variance of 0.336 and a standard deviation of 0.580. The high level of consistency is indicated by the variable's coefficient of variation, which is 8.359 %.

According to the study, the maximum value for the Dependent variable **Pull factors (y_3)** is 7.720, and the minimum value is 5.360.The study's overall distribution was determined to have a mean of 6.613 with a variance of 0.333 and a standard deviation of 0.577. The high level of consistency is indicated by the variable's coefficient of variation, which is 8.730 %.

Table 6.2: Coefficient of Correlation (r): Perceived Benefits (y_1) vs. 16 Independent Variables

Sl.No.	Independent Variables	'r' Value	Remarks
1	Age (x_1)	0.008	
2	Marital Status (x_2)	-0.049	
3	Age at the time of migration(x_3)	0.516	**
4	Education (x_4)	-0.129	
5	Family Size (x_5)	0.038	
6	Land Holding (x_6)	0.030	
7	Social Participation (x_7)	-0.056	
8	Pattern of Migration (x_8)	0.001	
9	Family income (x_9)	0.500	**
10	No. of information *Source* for migration(x_{10})	0.234	*
11	Social support network at destination(x_{11})	0.072	
12	No. of income *Sources*(x_{12})	0.246	*
13	Frequency of visiting home (x_{13})	-0.580	**
14	Duration of stay at home(x_{14})	-0.556	**
15	Distance of Migration(x_{15})	0.112	
16	Duration of Migration(x_{16})	-0.051	

**Correlation is significant at the 0.01 level *Correlation is significant at the 0.05 level

Results

Table 6.2 presents coefficient of correlation between perceived benefits of migration (y_1) and 16 causable variables (x_1-x_{16}). It has been found from the study that age at the time of migration (x_3), family income (x_9), no. of information *Source* for migration (x_{10}), no. of income *Source*s (x_{12}) has recorded significant positive correlation and frequency of visiting home (x_{13}), duration of stay at home (x_{14}) has recorded significant but negative correlation with Perceived Benefits of migration (y_1).

Revelation

The age at the time of migration (x_3) has recorded a significant positive correlation implying that it impacts on how individuals perceive the benefits of migration. Older individuals may have more life experience and maturity, which can shape their perception of the benefits of migration. They may have a better understanding of the potential advantages, such as improved economic opportunities, access to better education or healthcare, or a higher quality of life. This increased experience and maturity can contribute to a more positive perception of the benefits associated with migration. On the other hand younger individuals, particularly those who migrate at a relatively early age, may have greater adaptability and resilience when facing the challenges of migration. They may be more open to new experiences, learning opportunities, and cultural integration, which can enhance their perception of the benefits of migration. Additionally, their ability to adapt to new environments and navigate unfamiliar situations may lead to a positive outlook on the advantages that migration can bring. **Family income** plays a crucial role in determining the economic opportunities available to individuals. Higher family income can provide individuals with better financial re*Source*s, enabling them to pursue migration as a means of accessing improved job prospects, higher wages, and overall economic advancement. Individuals from families with higher incomes may perceive migration as an opportunity to enhance their financial well-being and achieve a higher standard of living. Family income can influence an individual's perception of social mobility. Individuals from low-income families may view migration as a means to escape poverty, overcome social barriers, and achieve upward social mobility. They may perceive the benefits of migration in terms of improving their social status, gaining access to better healthcare, housing, and overall quality of life. In contrast, individuals from higher-income families may perceive migration as a means to maintain or enhance their existing social and economic status. **No. of information *Source* for migration** has recorded a significant positive correlation which implies that when individuals have access to a greater number of information *Source*s, they are exposed to a wider range of perspectives and experiences related to migration. This exposure to diverse viewpoints allows individuals to form a more comprehensive understanding of the potential benefits of migration. It is quite logical that having multiple income

*Source*s can contribute to a greater sense of economic stability and security. When individuals rely on a single *Source* of income, they may perceive migration as an opportunity to diversify their income streams. By migrating to a new location, they may anticipate the potential to access additional job opportunities or establish new income *Source*s.

The variable **Frequency of visiting home** has recorded a strong but negative correlation. It is evident that individuals who migrate often have limited opportunities to visit their home place due to constraints such as tenure, time limitations, and negotiability regarding the frequency of visits.

Similarly the variable **Duration of stayat home** has also recorded a significant but negative correlation which implies that individuals or migrants may have limited opportunities to fully immerse themselves in the new culture, build social connections, or engage in activities that contribute to a sense of belonging. It is obvious that building a new life in a different location often requires consistent effort, such as finding employment, pursuing education, or developing social networks. By prolonging stays at home, individuals may experience disruptions in their routine and progress towards their goals, which can negatively impact their perception of the benefits of migration.

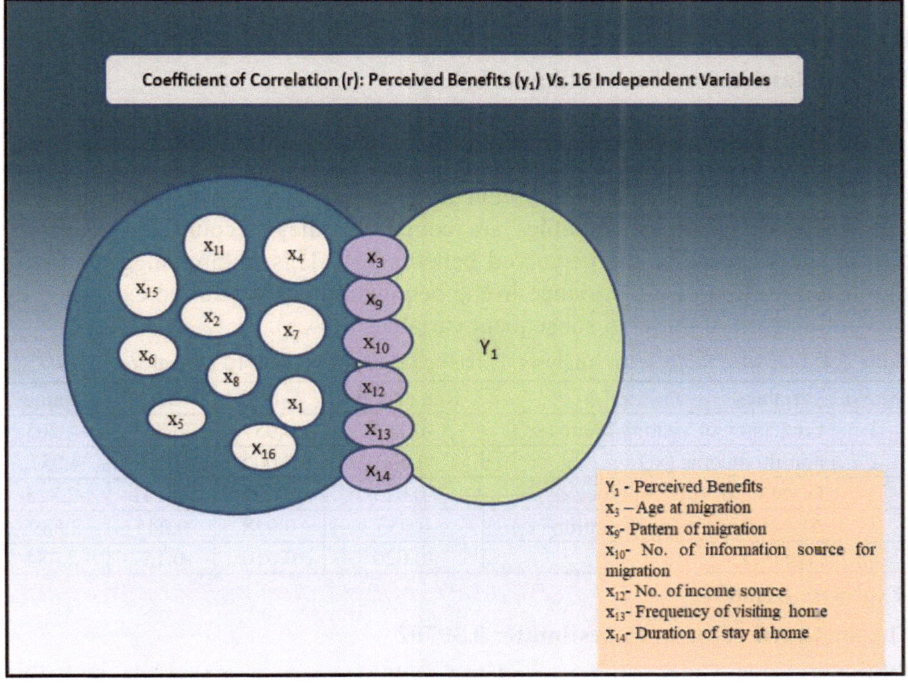

Fig. 6.1:

Table 6.3: Multiple Regression Analysis: Perceived Benefits (y_1) vs. 16 Causal Variables

Sl.No.	Variables	Reg. Coef. B	S.E. B	Beta	t Value
1	Age (x_1)	-0.027	0.023	-0.207	-1.156
2	Marital Status (x_2)	0.002	0.022	0.011	0.081
3	Age at the time of migration(x_3)	0.056	0.022	0.260	2.566
4	Education (x_4)	-0.015	0.014	-0.094	-1.086
5	Family Size (x_5)	0.015	0.049	0.028	0.303
6	Land Holding (x_6)	0.098	0.213	0.041	0.458
7	Social Participation (x_7)	-0.027	0.054	-0.041	-0.498
8	Pattern of Migration (x_8)	0.067	0.038	0.141	1.755
9	Family income (x_9)	0.000	0.000	0.376	3.728
10	No. of information *Source* for migration(x_{10})	0.033	0.047	0.067	0.704
11	Social support network at destination(x_{11})	-0.018	0.064	-0.023	-0.276
12	No. of income *Sources*(x_{12})	-0.111	0.104	-0.115	-1.068
13	Frequency of visiting home (x_{13})	-0.242	0.064	-0.353	-3.786
14	Duration of stay at home(x_{14})	-0.013	0.005	-0.244	-2.596
15	Distance of Migration(x_{15})	0.000	0.000	-0.015	-0.189
16	Duration of Migration(x_{16})	0.014	0.020	0.096	0.686

R square: 65.60%

The standard error of the estimate: 0.40814.

Results

In Table 6.3, the multiple regression analysis examines the relationship between perceived benefits (y_1) and 16 subsequent variables (x_1-x_{16}). The results indicate that when all 16 causal variables are combined, they account for 65.60 per cent of the variance in the perceived benefits (y_1). This finding suggests that a significant portion of the variance in the perceived benefits can be explained by the combination of these 16 subsequent variables (x_1-x_{16}).

Table 6.4: Stepwise Regression Analysis: Perceived Benefits (y_1) vs. 16 Causal Variables

Sl.No	Variables	Reg.coef. B	S.E. B	Beta	t value
1	Frequency of visiting home (x_{13})	-0.241	0.057	-0.351	-4.203
2	Family income (x_9)	0.000	0.000	0.309	4.057
3	Duration of stay at home (x_{14})	-0.013	0.005	-0.241	-2.825
4	Age at the time of migration (x_3)	0.053	0.018	0.245	2.889
5	Age (x_1)	-0.020	0.010	-0.157	-2.084

R square: 61.80%

The standard error of the estimate: 0.39702

Table 6.4 displays the complete model of multiple regression analysis, revealing that the collective influence of 5 causal variables *i.e.*Frequency of visiting home (x_{13}), Family income (x_9), Duration of stay at home (x_{14}), Age at the time of migration (x_3), Age (x_1), accounts for 61.80 percent of the variance embedded in the consequent variable, perceived benefits (y_1). The 16 causal variables collectively have explained 65.60 per cent but out of those these 5 variables together have explained 61.80 per cent, which strongly implies that these 5 variables play extremely significant role in terms of their contribution.

Perceived benefits is related to the frequency of visiting home since people tend to often choose to migrate to a favourable destination i.e. near destination. It is obvious that family income acts as a sole driver and motivation for accessing improved job prospects, higher wages, and overall economic advancement. Duration of stay at home is inversely related meaning that individuals or migrants may have limited opportunities to fully immerse themselves in the new culture, build social connections, or engage in activities that contribute to a sense of belonging or they are unsure for when will they get the opportunity to spend time in their native place and with their loved ones. Age at the time of migration explains that older individuals have increased experience and maturity and can contribute to a more positive perception of the benefits associated with migration. On the other hand younger individuals, particularly those who migrate at a relatively early age, may have greater adaptability and resilience when facing the challenges of migration.

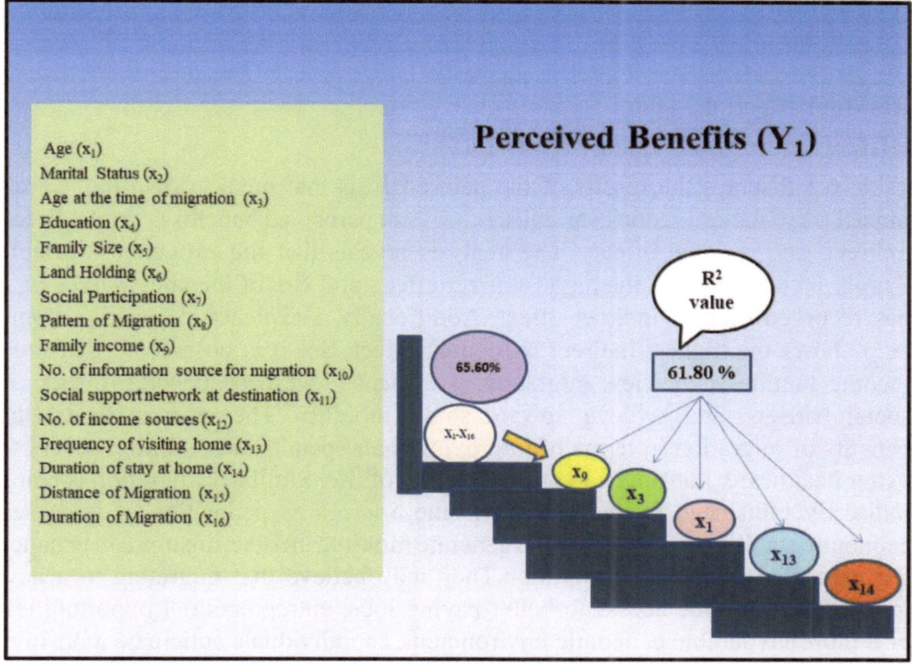

Fig. 6.2:

Table 6.5: Path Analysis: Decomposition of Total Effect into Direct, Indirect and Residual Effect: Y_1

Sl. No	Variables	Total Effect	Direct Effect	Indirect Effect	Highest Indirect Effect
1	Age (x_1)	0.008	-0.205	0.213	0.073 (x_{16})
2	Marital Status (x_2)	-0.049	0.012	-0.061	-0.169 (x_1)
3	Age at the time of migration(x_3)	0.516	0.260	0.256	0.122 (x_{13})
4	Education (x_4)	-0.129	-0.093	-0.036	0.065 (x_9)
5	Family Size (x_5)	0.038	0.029	0.009	-0.032 (x_{14})
6	Land Holding (x_6)	0.030	0.042	-0.012	0.032 (x_1)
7	Social Participation (x_7)	-0.056	-0.042	-0.014	-0.045 (x_1)
8	Pattern of Migration (x_8)	0.001	0.140	-0.139	-0.080 (x_{13})
9	Family income (x_9)	0.500	**0.375**	0.125	0.080 (x_3)
10	No. of information *Source* for migration(x_{10})	0.234	0.068	0.166	0.074 (x_{13})
11	Social support network at destination(x_{11})	0.072	-0.023	0.095	0.075 (x_9)
12	No. of income *Sources*(x_{12})	0.246	-0.114	**0.360**	0.209 (x_9)
13	Frequency of visiting home (x_{13})	**-0.580**	-0.352	-0.228	-0.114 (x_{14})
14	Duration of stay at home(x_{14})	-0.556	-0.245	-0.311	-0.164 (x_{13})
15	Distance of Migration(x_{15})	0.112	-0.015	0.127	0.036 (x_{13})
16	Duration of Migration(x_{16})	-0.051	0.095	-0.146	-0.159 (x_1)

Residual effect: 0.34406

HIIE :Frequency of visiting home(x_{13})(5).

Table 6.5 displays the results of the path analysis that breaks down the overall impact (r) of the exogenous variables (x_1-x_{16}) on perceived benefits (y_1) into direct, indirect, and residual effects. The analysis reveals that the exogenous variable Family income (x_9) has the highest direct effect, and No. of income *Source*s (x_{12}) has recorded highest indirect effect. Additionally, Frequency of visiting home (x_{13}) shows the highest indirect individual effect. So, it is observable that low-income families may view migration as a means to escape poverty, overcome social barriers, and achieve upward social mobility. They may perceive the benefits of migration in terms of improving their social status, gaining access to better healthcare, housing, and overall quality of life. Similarly migration is being indirectly influenced by the no. of income *Source*s as perception of increased economic stability and the ability to generate multiple income streams can enhance the perceived benefits of migration.They may believe that migrating to a new location can provide access to better-paying jobs, entrepreneurial opportunities, or a more favourable economic environment, as individuals anticipate a positive impact on their financial well-being. Frequency of visiting home implies that the migrants limited opportunities to visit their home place due to constraints such as tenure, time limitations, and negotiability regarding the frequency of visits. So they opt to choose the destination of migration keeping in mind the availability of visiting their native place frequently.The residual effect, accounting for 0.34406, implies that there is 34.40 per cent portion of the variance embedded in perceived benefits which cannot be explained by the combination of the 16 exogenous variables.

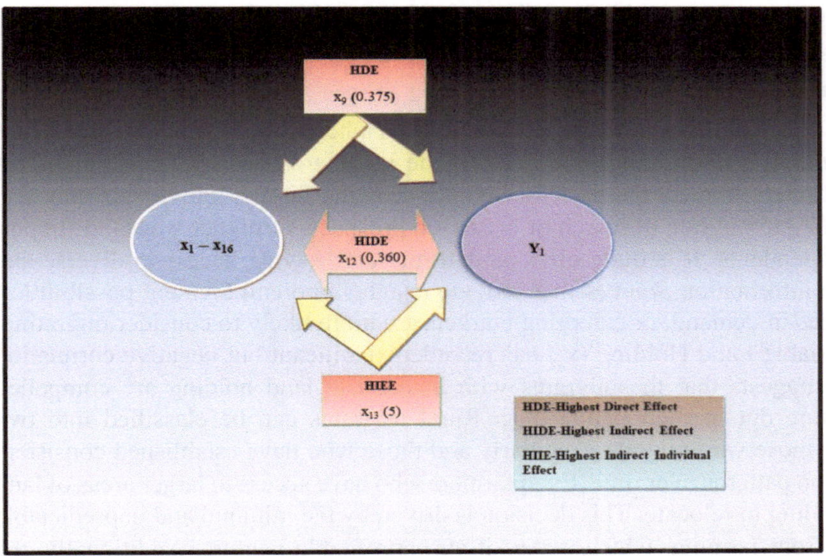

Fig. 6.3:

Table 6.6: Coefficient of Correlation (r): Push Factors (y_2) vs. 16 Independent Variables

Sl.No.	Independent Variables	'r' Value	Remarks
1	Age (x_1)	0.003	
2	Marital Status (x_2)	-0.067	
3	Age at the time of migration(x_3)	0.029	
4	Education (x_4)	-0.026	
5	Family Size (x_5)	0.571	**
6	Land Holding (x_6)	-0.571	**
7	Social Participation (x_7)	-0.010	
8	Pattern of Migration (x_8)	-0.019	
9	Family income (x_9)	0.030	
10	No. of information *Source* for migration(x_{10})	0.590	**
11	Social support network at destination(x_{11})	-0.157	
12	No. of income *Sources*(x_{12})	0.095	
13	Frequency of visiting home (x_{13})	-0.137	
14	Duration of stay at home(x_{14})	-0.024	
15	Distance of Migration(x_{15})	-0.145	
16	Duration of Migration(x_{16})	-0.079	

**Correlation is significant at the 0.01 level

*Correlation is significant at the 0.05 level

Results

Table 6.6 presents coefficient of correlation between push factors (y_2) and 16 causable variables (x_1-x_{16}). It has been found from the study that family size (x_5) and no. of information *Source* for migration (x_{10}) has recorded significant positive correlation and Land Holding (x_6) has recorded significant but negative correlation with Push factors (y_2).

Revelation

Family size can increase the financial strain on a household, especially in regions with limited economic opportunities. If the income earned by family members is insufficient to meet the basic needs of a growing family, individuals may feel compelled to migrate in search of better job prospects or higher wages to support their dependents. It is quite obvious that migrants having access to diverse and reliable information *Source*s that provide insights into employment possibilities, industries in demand, or emerging markets are more likely to consider migrating. The variable Land Holding (x_6) has recorded significant but negative correlation which suggests that the migrants with inadequate land holding are compelled to migrate out of their native place.Rural migrants can be classified into two groups: those who migrate irregularly and those who have established consistent migration patterns over time. Even farmers who have access to larger areas of land are deciding to relocate. This decision is driven by the minimal and unpredictable profits from farming, which makes it an unprofitable venture. Additionally, the desire for a more stable and respected way of life in a different location plays a role. These individuals aim to provide better support for their families, including education and healthcare for their spouses, parents, and children, while living in a remote area.

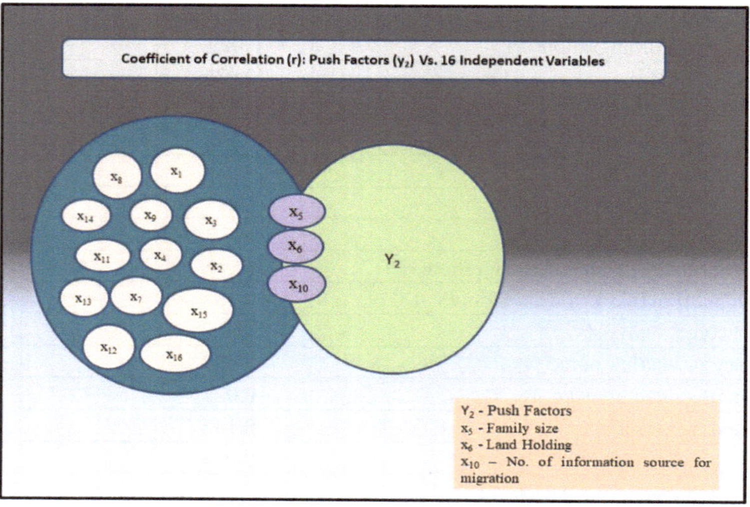

Fig. 6.4:

Table 6.7: Multiple Regression Analysis : Push factors (y_2) vs. 16 Causal Variables

Sl. No.	Variables	Reg. Coef. B	S.E. B	Beta	t Value
1	Age (x_1)	0.036	0.023	0.296	1.552
2	Marital Status (x_2)	-0.024	0.022	-0.166	-1.112
3	Age at the time of migration(x_3)	-0.023	0.022	-0.116	-1.072
4	Education (x_4)	0.005	0.014	0.033	0.356
5	Family Size (x_5)	0.170	0.049	0.344	3.467
6	Land Holding (x_6)	-0.671	0.212	-0.305	-3.163
7	Social Participation (x_7)	0.025	0.054	0.041	0.460
8	Pattern of Migration (x_8)	-0.032	0.038	-0.073	-0.856
9	Family income (x_9)	0.000	0.000	0.037	0.345
10	No. of information *Source* for migration(x_{10})	0.126	0.046	0.277	2.719
11	Social support network at destination(x_{11})	-0.072	0.064	-0.102	-1.132
12	No. of income *Sources*(x_{12})	-0.041	0.103	-0.045	-0.392
13	Frequency of visiting home (x_{13})	-0.094	0.064	-0.147	-1.476
14	Duration of stay at home (x_{14})	0.001	0.005	0.011	0.107
15	Distance of Migration(x_{15})	0.000	0.000	-0.109	-1.266
16	Duration of Migration(x_{16})	-0.030	0.020	-0.225	-1.505

R square: 60.90%

The standard error of the estimate: 0.40593.

Results

In Table 6.7, the multiple regression analysis examines the functional relationship between push factors (y_2)and 16 subsequent variables (x_1-x_{16}). The results indicate that when all 16 causal variables are combined, they account for 60.90 per cent of the variance in thepush factors (y_2). This finding suggests that a significant portion of the variance in the perceived benefits can be explained by the combination of these 16 subsequent variables (x_1-x_{16}).

Table 6.8: Stepwise Regression Analysis: Push factors (y_2) vs. 16 Causal Variables

Sl.No	Variables	Reg.coef.B	S.E. B	Beta	t value
1	No. of information *Source* for migration(x_{10})	0.143	0.042	0.313	3.433
2	Land Holding (x_6)	-0.723	0.191	-0.329	-3.780
3	Family Size (x_5)	0.151	0.044	0.307	3.443

R square: 54.70%

The standard error of the estimate: 0.39781

Table 6.8 displays the complete model of multiple regression analysis, revealing that the collective influence of 3 causal variables *i.e.* No. of information *Source* for migration(x_{10}), Land Holding (x_6), Family Size (x_5), accounts for 54.70 percent of the variance embedded in the consequent variable, push factors (y_2).The 16 causal variables collectively have explained 60.90 per cent but out of those these 3 variables together have explained 54.70 per cent, which strongly implies that these 3 variables play extremely significant role in terms of their contribution.

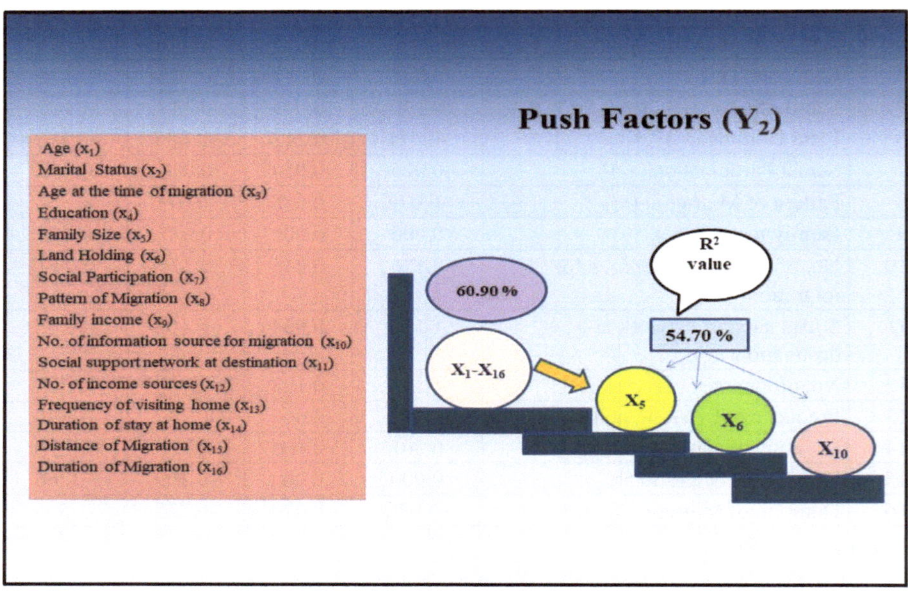

Fig. 6.5:

Table 6.9: Path Analysis: Decomposition of Total Effect into Direct, Indirect and Residual Effect: Y_2

Sl. No	Variables	Total Effect	Direct Effect	Indirect Effect	Highest Indirect Effect
1	Age (x_1)	0.003	0.296	-0.293	-0.172 (x_{16})
2	Marital Status (x_2)	-0.067	-0.168	0.101	0.243 (x_1)
3	Age at the time of migration(x_3)	0.029	-0.115	0.144	0.082 (x_1)
4	Education (x_4)	-0.026	0.032	-0.058	-0.033 (x_{11})
5	Family Size (x_5)	0.571	**0.343**	0.228	0.128 (x_{10})
6	Land Holding (x_6)	-0.571	-0.307	-0.264	-0.126 (x_5)
7	Social Participation (x_7)	-0.010	0.040	-0.050	0.064 (x_1)
8	Pattern of Migration (x_8)	-0.019	-0.073	0.054	0.071 (x_5)
9	Family income (x_9)	0.030	0.038	-0.008	0.04 (x_{10})
10	No. of information *Source* for migration(x_{10})	**0.590**	0.276	**0.314**	0.159 (x_5)
11	Social support network at destination(x_{11})	-0.157	-0.102	-0.055	-0.038 (x_6)
12	No. of income *Sources*(x_{12})	0.095	-0.044	0.139	0.075 (x_1)

Sl. No	Variables	Total Effect	Direct Effect	Indirect Effect	Highest Indirect Effect
13	Frequency of visiting home (x_{13})	-0.137	-0.147	0.010	-0.058 (x_{10})
14	Duration of stay at home(x_{14})	-0.024	0.011	-0.035	-0.069 (x_{13})
15	Distance of Migration(x_{15})	-0.145	-0.109	-0.036	-0.028 (x_{10})
16	Duration of Migration(x_{16})	-0.079	-0.223	0.144	0.228 (x_1)

Residual effect: 0.39099

HIEE: Age (x_1) (5)

Table 6.9 displays the results of the path analysis that breaks down the overall impact (r) of the exogenous variables (x_1-x_{16}) on push factors (y_2) into direct, indirect, and residual effects. The analysis reveals that the exogenous variable Family Size (x_5) has the highest direct effect and No. of information *Source* for migration (x_{10}) has recorded highest indirect effect. Additionally, Age (x_1) shows the highest indirect individual effect.

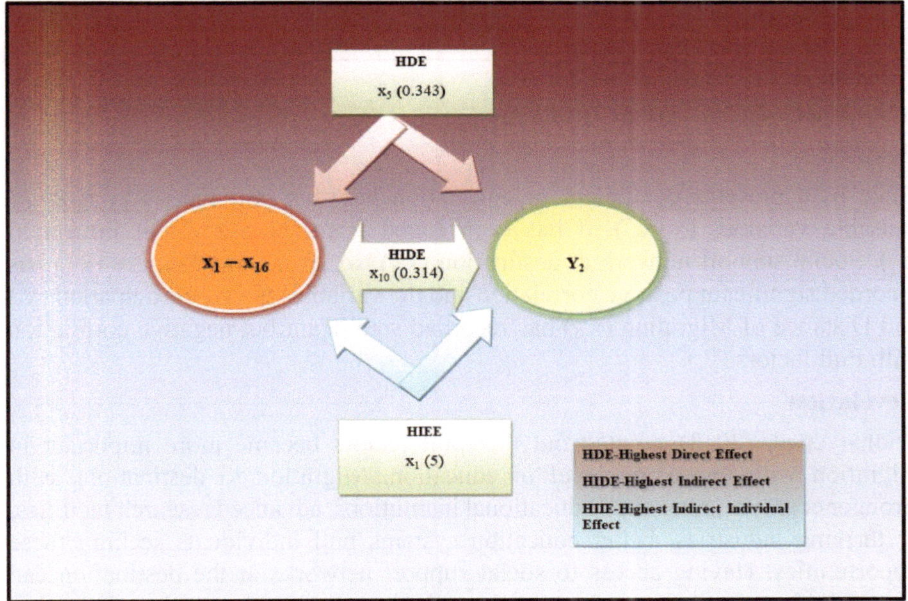

Fig. 6.6:

Table 6.10: Coefficient of Correlation (r): Pull factors (y_3) vs. 16 Independent Variables

Sl. No.	Independent Variables	'r' Value	Remarks
1	Age (x_1)	0.089	
2	Marital Status (x_2)	0.097	
3	Age at the time of migration(x_3)	-0.028	
4	Education (x_4)	0.305	**
5	Family Size (x_5)	-0.008	
6	Land Holding (x_6)	0.065	
7	Social Participation (x_7)	-0.423	**
8	Pattern of Migration (x_8)	-0.048	
9	Family income (x_9)	0.087	
10	No. of information *Source* for migration(x_{10})	0.084	
11	Social support network at destination(x_{11})	0.396	**
12	No. of income *Sources*(x_{12})	0.337	**
13	Frequency of visiting home (x_{13})	0.074	
14	Duration of stay at home(x_{14})	-0.036	
15	Distance of Migration(x_{15})	-0.310	**
16	Duration of Migration(x_{16})	0.063	

**Correlation is significant at the 0.01 level

*Correlation is significant at the 0.05 level

Results

Table 6.10 presents coefficient of correlation between pull factors (y_3)and 16 causable variables (x_1-x_{16}). It has been found from the study that Education (x_4), Social support network at destination (x_{11}),No. of income *Sources* (x_{12}) has recorded significant positive correlation and the variables Social Participation (x_7) and Distance of Migration (x_{15}) has recorded significant but negative correlation with Pull factors (y_3).

Revelation

Sridhar et al, (2013) pointed out that pull factors became more important in migration with increasing level of education. Migration to destinations with stronger economy, renowned educational institutions, advanced research facilities, or thriving industries, better education systems pull individuals seeking these opportunities. Having access to social support networks at the destination can be a significant pull factor for migrants. Having significant amount of existing communities, friends, or family members who can provide assistance, guidance and emotional support makes the process of migration less daunting. It is quite evident that presence of multiple income *Sources* at the destination can offer migrants the opportunity to earn a higher income. Increased earning potential acts as a significant pull factor for individuals seeking to improve their life standard, accumulate savings, financially support their family. The variable Social participation has recorded a significant but negative correlation which implies that active social participation develops a strong social network and support systems within a community which provide individuals with emotional, financial, practical and social support. When a migrant has strong support within their community, they are less inclined to migrate. Additionally, the pull factors of migration related to employment, higher wages, and improved job prospects are often attractive when they are within a reachable distance. Proximity can facilitate connections with fellow expatriates or members of the same ethnic or cultural community which can help in navigating the new environment.

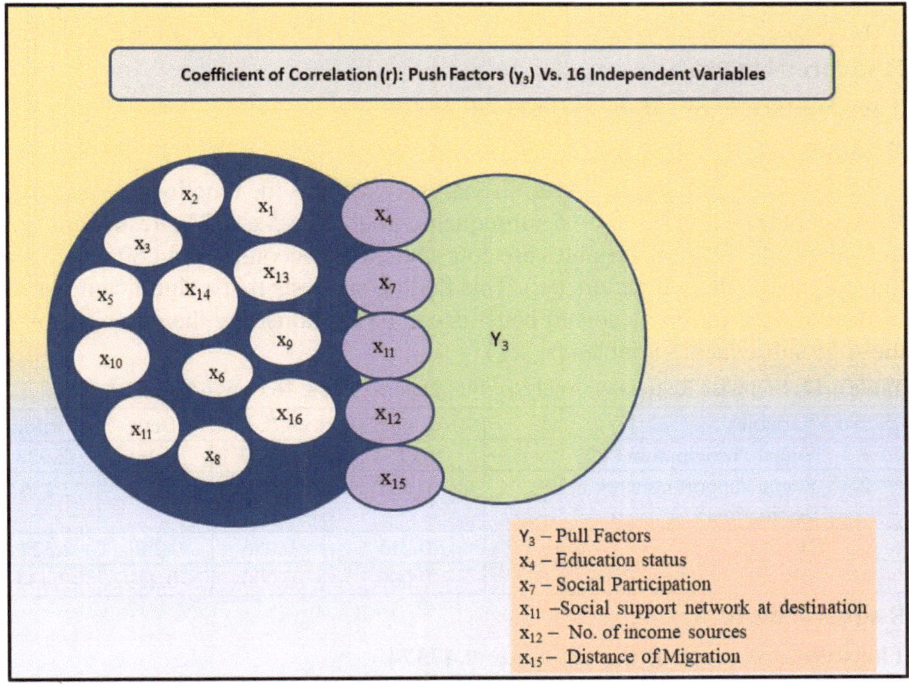

Fig. 6.7:

Table 6.11: Multiple Regression Analysis : Pull factors (y_3) vs. 16 Causal Variables

Sl.No.	Variables	Reg. Coef. B	S.E. B	Beta	t Value
1	Age (x_1)	-0.016	0.027	-0.133	-0.585
2	Marital Status (x_2)	0.026	0.026	0.181	1.014
3	Age at the time of migration(x_3)	0.007	0.026	0.035	0.273
4	Education (x_4)	0.022	0.017	0.145	1.317
5	Family Size (x_5)	-0.001	0.058	-0.003	-0.024
6	Land Holding (x_6)	0.114	0.251	0.052	0.452
7	Social Participation (x_7)	-0.179	0.064	-0.295	-2.793
8	Pattern of Migration (x_8)	-0.046	0.045	-0.104	-1.021
9	Family income (x_9)	0.000	0.000	-0.167	-1.303

Sl.No.	Variables	Reg. Coef. B	S.E. B	Beta	t Value
10	No. of information *Source* for migration(x_{10})	0.014	0.055	0.031	0.258
11	Social support network at destination(x_{11})	0.136	0.076	0.193	1.801
12	No. of income *Sources*(x_{12})	0.316	0.123	0.353	2.579
13	Frequency of visiting home (x_{13})	0.126	0.075	0.198	1.673
14	Duration of stay at home(x_{14})	-0.005	0.006	-0.096	-0.802
15	Distance of Migration(x_{15})	0.000	0.000	-0.177	-1.725
16	Duration of Migration(x_{16})	0.016	0.024	0.120	0.676

R square: 44.60%

The standard error of the estimate: 0.48113.

Results

In Table 6.11, the multiple regression analysis examines the functional relationship between pull factors (y_3)and 16 subsequent variables (x_1-x_{16}). The results indicate that when all 16 causal variables are combined, they account for 44.60 per cent of the variance in thepull factors (y_3). This finding suggests that a significant portion of the variance in the perceived benefits can be explained by the combination of these 16 subsequent variables (x_1-x_{16}).

Table 6.12: Stepwise Regression Analysis: Pull factors (y_3) vs. 16 Causal Variables

Sl.No	Variables	Reg.coef.B	S.E. B	Beta	t value
1	Social Participation (x_7)	-0.174	0.059	-0.286	-2.923
2	Social support network at destination(x_{11})	0.157	0.070	0.222	2.236
3	No of income *Sources* (x_{12})	0.215	0.085	0.240	2.527
4	Distance of Migration(x_{15})	0.000	0.000	-0.229	-2.434

R square: 36.10%

The standard error of the estimate: 0.47374.

Table 6.12 displays the complete model of multiple regression analysis, revealing that the collective influence of 4 causal variables *i.e.* Social Participation (x_7), Social support network at destination (x_{11}), No of income *Sources* (x_{12}), Distance of Migration(x_{15}), accounts for 36.10 percent of the variance embedded in the consequent variable, pull factors (y_3).The 16 causal variables collectively have explained 44.60 per cent but out of those these 4 variables together have explained 36.10 per cent, which implies that these 4 variables play slightly significant role in terms of their contribution.

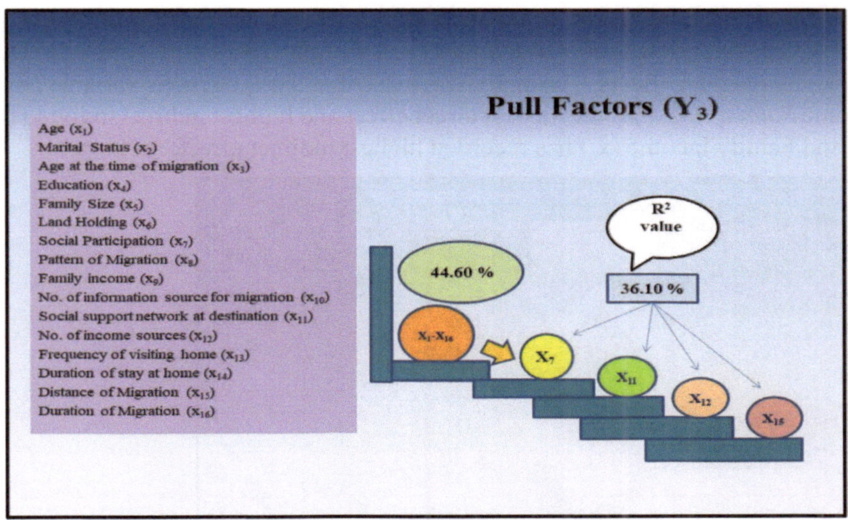

Fig. 6.8:

Table 6.13: Path Analysis: Decomposition of Total Effect into Direct, Indirect and Residual Effect: Y_3

Sl.No	Variables	Total Effect	Direct Effect	Indirect Effect	Highest Indirect Effect
1	Age (x_1)	0.089	-0.134	0.223	0.151 (x_2)
2	Marital Status (x_2)	0.097	0.183	-0.086	-0.111 (x_1)
3	Age at the time of migration(x_3)	-0.028	0.035	-0.063	0.081 (x_{12})
4	Education (x_4)	0.305	0.145	0.160	0.061 (x_{11})
5	Family Size (x_5)	-0.008	-0.003	-0.005	0.068 (x_{12})
6	Land Holding (x_6)	0.065	0.051	0.014	-0.029 (x_2)
7	Social Participation (x_7)	**-0.423**	-0.295	-0.128	-0.059 (x_{11})
8	Pattern of Migration (x_8)	-0.048	-0.104	0.056	0.053 (x_{12})
9	Family income (x_9)	0.087	-0.167	**0.254**	0.197 (x_{12})
10	No. of information *Source* for migration(x_{10})	0.084	0.031	0.053	0.094 (x_{12})
11	Social support network at destination(x_{11})	0.396	0.194	0.202	0.091 (x_7)
12	No. of income *Sources*(x_{12})	0.337	**0.353**	-0.016	-0.093 (x_9)
13	Frequency of visiting home (x_{13})	0.074	0.197	-0.123	-0.07 (x_{12})
14	Duration of stay at home(x_{14})	-0.036	-0.097	0.061	0.093 (x_{13})
15	Distance of Migration(x_{15})	-0.310	-0.177	-0.133	-0.043 (x_7)
16	Duration of Migration(x_{16})	0.063	0.119	-0.056	0.121 (x_2)

Residual effect: 0.55362

HIEE: No. of income *Source*s (x_{12}) (6)

Table 6.13 displays the results of the path analysis that breaks down the overall impact (r) of the exogenous variables (x_1-x_{16}) on pull factors (y_3) into direct, indirect, and residual effects. The analysis reveals that the exogenous variable No. of income *Source*s (x_{12}) has the highest direct effect and highest indirect individual effect and Family income (x_9) has recorded highest indirect effect.

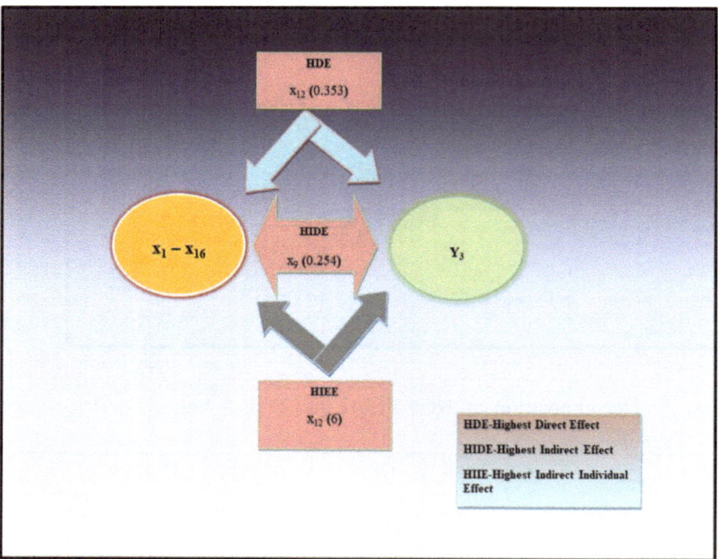

Fig. 6.9:

Table 6.14: Factor Analysis: Strategic Conglomeration of Variables into Factor

Factors	Variables	Factor loading	% of variance	Cumulative %	Factor renamed
F-1	Age at the time of migration(x_3) Family income(x_9) No. of information *Source* for migration(x_{10}) No. of income *Sources*(x_{12}) Frequency of visiting home(x_{13}) Duration of stay at home(x_{14}) Perceived Benefits (y_1)	3.25 2.679 2.493 2.218 1.391 0.975 0.894	17.105	17.105	Family ecology
F-2	Age (x_1) Marital Status (x_2) Social Participation (x_7) Duration of Migration (x_{16}) Push Factors (y_2)	0.795 0.698 0.597 0.561 0.506	14.100	31.205	Social ecology of Migration

F-3	Family Size (x_5) Land Holding (x_6) Social support network at destination (x_{11})	0.433 0.405 0.352	13.122	44.327	Family re*Source*
F-4	Education (x_4) Distance of Migration (x_{15}) Pull Factors (y_3)	0.284 0.203 0.174	11.673	56.001	Migration drivers
F-5	Pattern of Migration (x_8)	0.094	7.321	63.321	

Table 6.14illustrates the factor analysis by highlighting the strategic conglomeration of variables into sub factors based on factor loading.

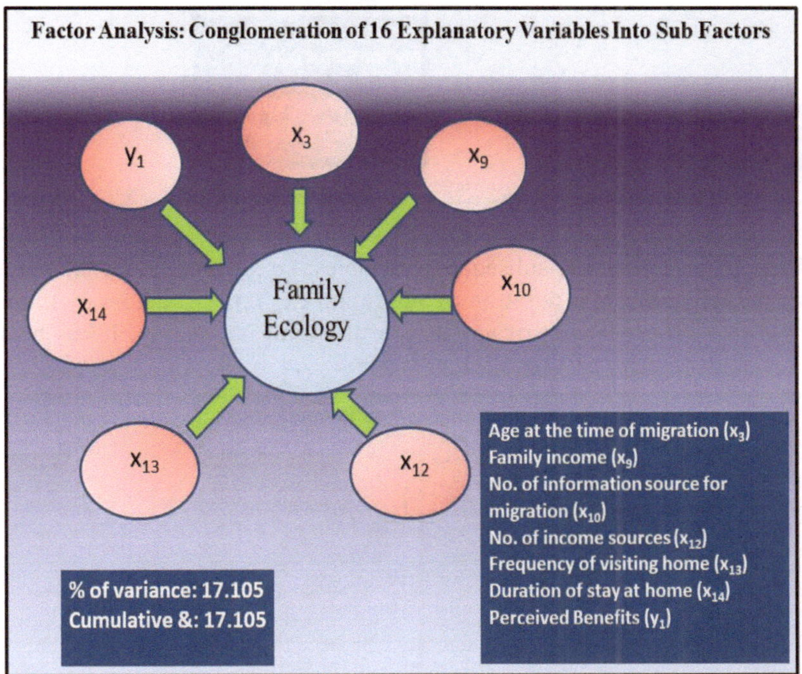

Fig. 6.10:

Factor I has incorporated the subsequent variables, *i.e.*,age at the time of migration (x_3), family income (x_9), no. of information *Source* for migration (x_{10}), no. of income *Source*s (x_{12}), frequency of visiting home (x_{13}), duration of stay at home (x_{14}), perceived benefits (y_1); and has contributed 17.105 per cent of variance of the total interaction. It has been renamed as Family ecology.

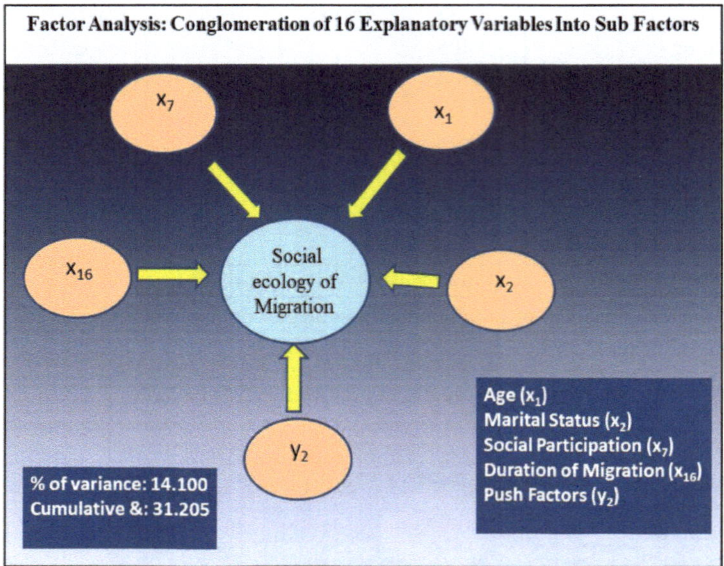

Fig. 6.11:

Factor II has incorporated the subsequent variables, *i.e.*, age (x_1), marital status (x_2), social participation (x_7), duration of migration (x_{16}), push factors (y_2); and has contributed 14.100 per cent of variance of the total interaction. It has been renamed as Social ecology of Migration.

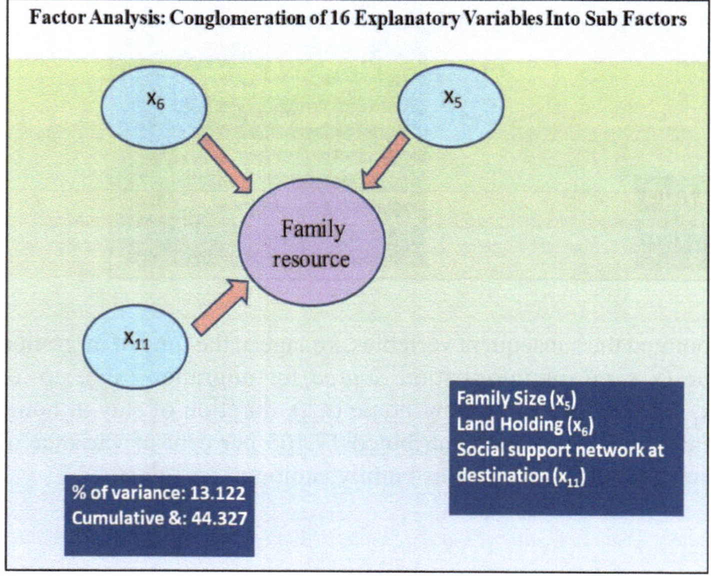

Fig. 6.12:

Factor III has incorporated the subsequent variables, *i.e.*, family size (x_5), land holding (x_6), social support network at destination (x_{11}); and has contributed 13.122 per cent of variance of the total interaction. It has been renamed as Family re*Source*.

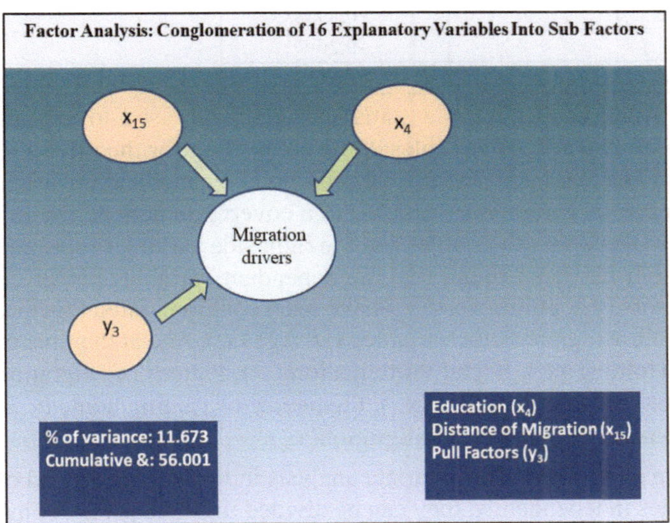

Fig. 6.13:

Factor IV has incorporated the subsequent variables, *i.e.*, education (x_4), distance of migration (x_{15}), pull factors (y_3); and has contributed 11.673 per cent of variance of the total interaction. It has been renamed as Migration drivers.

Table 6.15: Canonical Covariate Analysis: Interaction between Left Side Dependent and Right Side Independent Variables

Dependent Variable	**Coefficients**
Perceived Benefits(y_1)	-0.749
Push Factors (y_2)	-0.864
Pull Factors (y_3)	-0.805

Independent Variable	**Coefficients**
Age (x_1)	-0.627
Marital Status (x_2)	0.425
Age at the time of migration (x_3)	-0.133
Education (x_4)	0.043
Family Size (x_5)	0.178
Land Holding (x_6)	-0.359
Social Participation (x_7)	-0.285
Pattern of Migration (x_8)	-0.003
Family income (x_9)	0.161
No. of information *Source* for migration(x_{10})	0.244

Independent Variable	Coefficients
Social support network at destination(x_{11})	-0.051
No. of income *Sources*(x_{12})	0.240
Frequency of visiting home (x_{13})	-0.307
Duration of stay at home(x_{14})	-0.497
Distance of Migration(x_{15})	-0.269
Duration of Migration(x_{16})	0.072

Canonical Covariate Analysis (CCA) is a statistical technique used to explore the relationship between two sets of variables while controlling for the effects of other variables known as covariates. Table 6.14 presents the canonical covariate analysis between two sets of variables (Y and X) in a covert interaction. The left side variables represent the Y set of variables and the right side variables represent the X set of variables. It has been found that the dependent variables perceived benefits (y_1), push factors (y_2), pull factors (y_3) have been conglomerated together and impacted by the following predictor variables viz. Age (x_1), Age at the time of migration (x_3), Land Holding (x_6), Social Participation (x_7), Pattern of Migration (x_8), Social support network at destination (x_{11}), Frequency of visiting home (x_{13}), Duration of stay at home(x_{14}), Distance of Migration(x_{15}).

The comprehensive range of canonical covariate analysis indicates that instead of treating the 'y' variables independently, they can be divided into groups. By doing so, we can effectively focus on the selected 'x' variables on the right side while considering the group of 'y' variables separately.

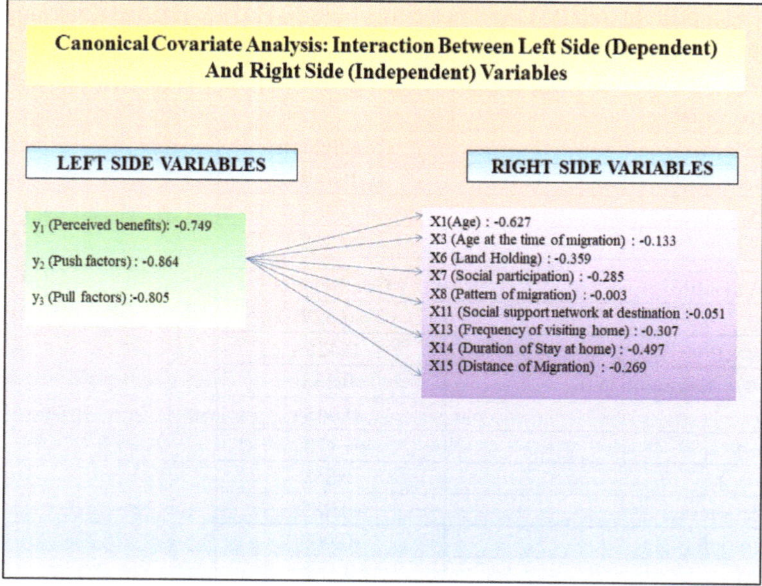

Fig. 6.14:

6.2 Shared Mental map of multiple stakeholders

Migration is a complex phenomenon that has influenced societies and human history for many years. In search of better prospects, safety, or a different way of life, it involves the movement of individuals or groups of people from one place to another, frequently across national or international lines. The causes of migration are numerous and varied, ranging from monetary factors and political unrest to alterations in the environment and social dynamics. Migration, along with fertility and mortality, is a crucial indicator of shifts in the social and economic domains. It encompasses the movement of people between different regions within a country and across international borders. The historical roots of both types of migration in India can be traced back to the nineteenth century, if not even earlier (Chattopadhyaya, 1987).

Here, we tried to study Migration in Darjeeling region of West Bengal using *the MentalModeler* software. We have created mental models about migration through Fuzzy Cognitive Mapping in the research area. Then a single map was generated by conglomerating each stakeholders distinct cognitive maps (Fig. 6.15) to understand migration in nearby villages. The node property and network property of shared mental map were examined lastly we were able to isolate the drivers, ordinaries and receivers by examining the indegree, outdegree and centrality values of each element.

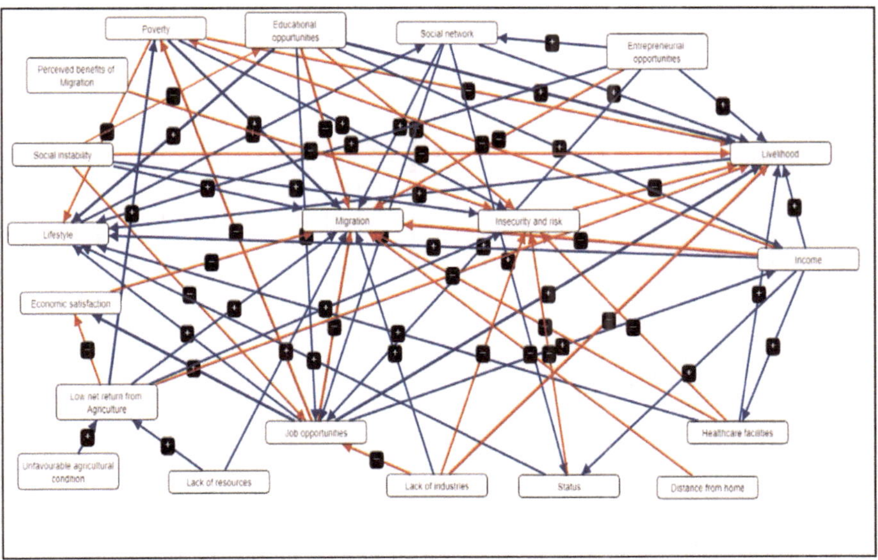

Fig. 6.15: Shared mental map of different stakeholders regarding migration.

From the viewpoint of each stakeholders a Fuzzy Cognitive map of Migration is shown in Fig.6.14. This one map combines all the separate maps made by various stakeholders. We obtain a clear picture because all the possible elements are present in this map. There are 20 elements and 60 connections between them.

Table 6.16: Selected determinants of Migration according to multiple stakeholders.

6.2.1 Network property

Total Component	Total Connection	Density	Connections/ Component	No. of Driver Component	No. of Receiver Component	No. of Ordinary Component	Complexity Score
20	60	0.1578947368	3	7	1	12	0.1428571429

6.2.2 Node property

Component	**Indegree**	**Outdegree**	**Centrality**	**Type**
Poverty	2.31	3.3	5.60	ordinary
Insecurity and risk	5.62	0.83	6.45	ordinary
Livelihood	8.61	0.74	9.35	ordinary
Job opportunities	3.48	4.74	8.22	ordinary
Educational opportunities	0.46	4.5	4.96	ordinary
Social network	1.41	2.60	4.01	ordinary
Entrepreneurial opportunities	0	3.61	3.62	driver
Healthcare facilities	0.64	2.77	3.41	ordinary
Lifestyle	6.12	0.73	6.85	ordinary
Social instability	0	3.37	3.37	driver
Status	1.47	1.57	3.04	ordinary
Income	1.05	4.09	5.14	ordinary
Low net return from Agriculture	1.55	3.49	5.04	ordinary
Economic satisfaction	1.5	0.65	2.15	ordinary
Perceived benefits of migration	0	0.52	0.52	driver
Lack of Industries	0	3.18	3.18	driver
Unfavourable agricultural condition	0	0.75	0.75	driver
Lack of re*Sources*	0	1.45	1.45	driver
Distance from home	0	0.7	0.7	driver
Migration	0.38	0	9.38	receiver

Results: If we compare Fig. 1 and Table 1 we can see that Social instability and Entrepreneurial opportunities have significant outdegree value with zero indegree value thus making them system drivers. So these two driver elements contribute significantly as drivers for migration in the study area.

Software detected Lack of industries, perceived benefits of migration, unfavourable agricultural conditions, lack of re*Source*s, distance from home in migration as drivers, but these micro elements should not be considered as drivers because of low outdegree value. So these micro elements are the other reasons that contribute in the migration process. On the other hand, the elements like Insecurity and risk, Livelihood, Job opportunities, Lifestyle all have remarkable indegree value with certain outdegree value and they are connected to other factors such as Educational opportunities, poverty, social instability, lack of industries, healthcare facilities, status, job opportunities, low net return from agriculture, perceived benefits of migration, entrepreneurial opportunities, social network, income, low net return from agriculture. These are the ordinary elements in migration.

Livelihood has the highest indegree and centrality value among ordinary elements because it is linked to many elements. Livelihood is a major component of migration that controls the process of migration. Job opportunities, income, low net return from agriculture and Poverty have significant outdegree value with certain indegree value. Other crucial elements like Social network and Status; which have significant indegree and outdegree value resulting in more centrality value.

Software detected migration as receiver of the system because it has significant indegree value, with zero outdegree value. As our study was focused on investigating the reasons and perception behind migration, so it has been found that migration was caused by some other elements. Hence, it has significant indegree with zero outdegree value.

6.2.3 Scenario analysis

We conducted scenario analysis Fig. 2 using the shared mental model of all the stakeholders. It was performed in order to generate detailed and more predictable scenario outcomes. We simulated 2 distinct scenarios using Mental Modeler's Scenario Module, taking into account the current situation and the viewpoint of stakeholders. Two scenarios were

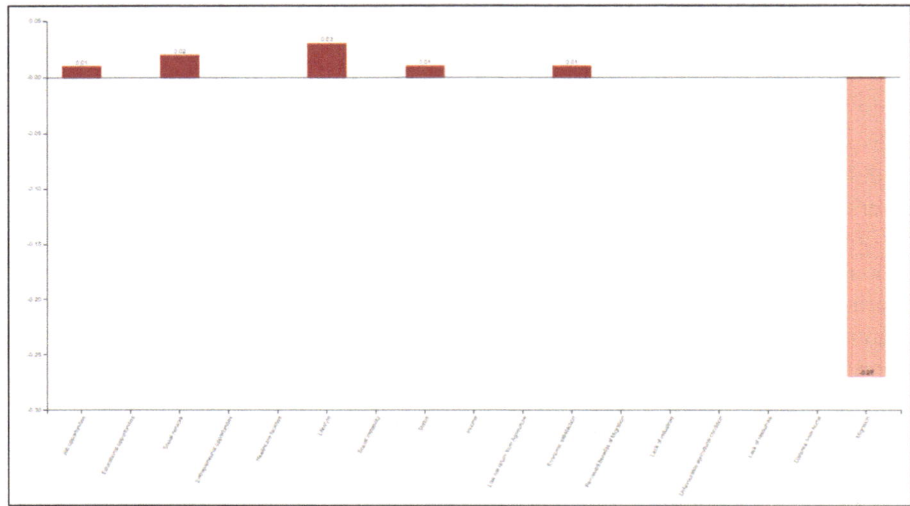

Fig. 6.16: Scenario 1: we reduced Poverty, Insecurity and risk and enhanced Livelihood

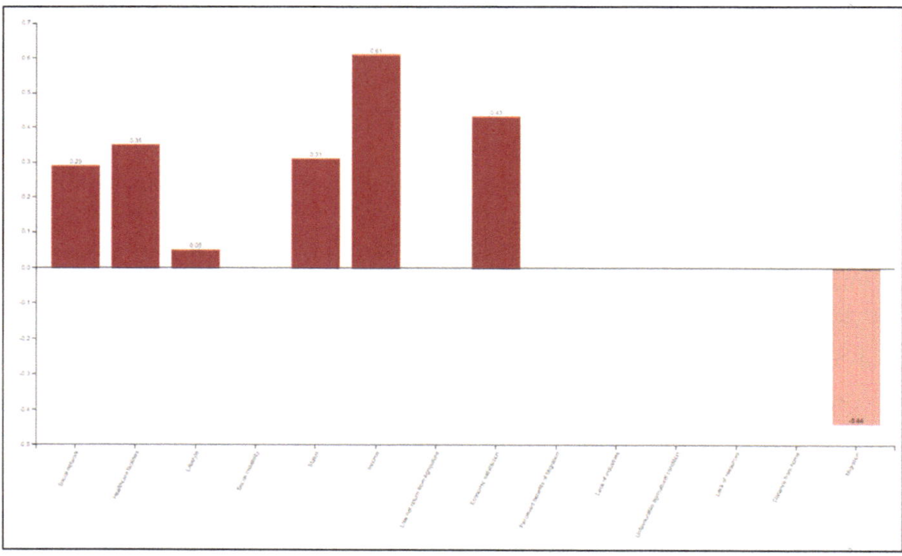

Fig. 6.17: Scenario 2: In addition to Scenario 1, we enhanced Job opportunities, Educational opportunities, and Entrepreneurial opportunities

Results

Either positive or negative impacts were observed on the elements of migration after simulation. Under scenario-1, Job opportunities, social network, lifestyle, status, and economic satisfaction were increased and migration showed a decline.

Under Scenario-2, income showed a noticeable increase and migration declined significantly. Additionally healthcare facilities also increased.

7

Epilogue

Migration is an integral part of civilization. Everyone is born in a location, wherein, for education he/she needs to migrate elsewhere and after completion of education, there would be further migration for jobs. During Covid-19 it was harshly depicted that more than 50 million migratory workers were forced to quit their job place and when the Covid-19 pandemic apocalypse was over they started re-migrating. Both Push and Pull factors are responsible for Migration. Pull factors incubates choices for migration ad push factors creates a compulsion for migration.Young people from all backgrounds are looking into different ways for rural and urban youth to get about in the town of Darjeeling. Urban adolescents are faced with downward mobility and endure disappointment in their hopes for professional professions and 'modern' lives, whereas rural kids enjoy certain types of upward mobility when they relocate into the metropolis (Brown, T. 2015). It is recognized that the main driving force behind migration is the pursuit of better job prospects, and the extent and direction of migration are predominantly influenced by employment opportunities (Khan, 1983). Furthermore, rural areas in Darjeeling have experienced a noticeable trend of male-selective migration in search of better opportunities in the past few decades. This migration primarily occurs towards urban environments within the district and major cities such as Delhi, Kolkata, Chennai, and Mumbai, to name a few important destinations. Several factors have contributed significantly to this phenomenon, including population pressure, degradation of common property re*Source*s like forests, long neglect of the region, economic and social backwardness, low agricultural productivity, and lack of opportunities. While this migration brings in much-needed income, it also diminishes the available workforce in the home villages. As a result, agricultural production has been severely disrupted and is ultimately declining. The outflow of labour force not only damages the subsistence economy but also discourages efforts towards self-reliance in the villages. Although the money earned through migration is sent back to the villages to purchase provisions, ecological constraints, inadequate manpower, and reliance on traditional low-tech agriculture prevent the local hill population from generating sufficient income to meet their basic needs. Consequently, the money circulates back to the plains to purchase provisions transported from there, creating a vicious cycle that leaves the rural people of Darjeeling Himalaya with minimal savings (Khawas, 2002).

Darjeeling hill areas are no exception to it. Due to shimmering crisis in Tea gardens and fallout of tourism, the young people have found migration a better choice for them. The surrounding agriculture has not been that remunerative. That is also the reasons for migration. Present-day Darjeeling has issues with traffic, pollution, and a loss of beautiful landscapes. Urban sprawl is a result of rising mobility and immigration. To accommodate the expanding population, especially the growing number of rural migrants who have been obliged to leave their homes due to declining rural jobs, much of Darjeeling's architectural legacy has vanished and many new projects have emerged (Ruchira and Timothy, 2015).

All kinds of migration are prevalent in Darjeeling, viz. Territorial, Occupational and Cultural migration as well.

The study reveals the need for framing resilient policy to generate ample scope for job creation, redressal of issues related to Tea garden management and making tourism industry more belligerent and job generating. The insecurity surrounding employment and the possibility of losing one's means of livelihood in the tea plantations are causing a new trend of people migrating to urban areas. As a consequence, new structures are being built on sloping land, which is leading to a shortage of space, increased pressure on available land, scarcity of water, and difficulties with drainage in the region. (Sarkar and Lama 1986)

This kind of researches can be replicated at different altitudes and terrains of ecosystem so that the geo- territorial features can be highlighted and elucidated properly. Already such type of researches has been carried out by Sankar Kr Acharya and DristikaJairu (2020); Altitude Extension: Socio-ecological dimension of Darjeeling tea garden.

If we want to keep Darjeeling, world's one of the most beautiful hill station smiling, we need to have a comprehensive plan and its execution; only romance and passion for Darjeeling are not good enough to ensure a preferred persistent development for the hill people primarily and the nation at last.

8.1 Limitations

1. The research was limited to a specific geographic area.
2. The study was constrained by limitations in time and re*Source*s.
3. The study relies on the subjective opinions of the participants, which may be influenced by their individual perspectives and biases, despite the researcher's efforts to gather them as objectively as possible.
4. It would have been advantageous to include a broader range of relevant variables in the study.
5. A suitable statistical model could have been employed to provide thematic support for the behavior and functioning of the system.
6. The selection of variables could have been expanded by incorporating additional variables that would have demonstrated resilience in domain interactions.

8.2 Future Scope of Research

1. A longitudinal study can be conducted to examine the dynamics of migration in Darjeeling over an extended period. This would help capture changes in the social ecology of migration and identify emerging patterns and trends.
2. Comparatively analyzing the social ecology of migration in Darjeeling with other regions or similar contexts. This would provide a broader understanding of the factors influencing migration and their specific implications for the local social ecology.
3. Investigate the socioeconomic factors that contribute to migration in Darjeeling. Explore how economic conditions, employment opportunities, and income disparities affect the decision to migrate and shape the social dynamics within migrant communities.
4. Explore the mental models and perceptions of migrants regarding their chosen destinations. Investigate how these mental models influence decision-making processes, adaptation strategies, and social interactions within migrant communities.
5. Examine the social consequences of migration in Darjeeling. Investigate how migration affects social networks, community cohesion, cultural practices, and identity formation, both among migrants and in the host community.
6. Assess the implications of migration on urban planning, infrastructure development, and public services in destination areas. Identify strategies to address the challenges posed by migration, including housing, transportation, healthcare, and education.
7. Investigate the relationship between climate change and migration in Darjeeling. Examine how environmental factors, such as changing weather patterns and natural disasters, interact with socioeconomic conditions to influence migration decisions and shape the social ecology.
8. Explore approaches to enhance community engagement and empower migrants in decision-making processes. Investigate how participatory methods and community-driven initiatives can promote inclusive development and address the social and economic needs of migrant populations.
9. Develop evidence-based policy recommendations to address the social ecological challenges associated with migration in Darjeeling. Provide insights on fostering sustainable development, social integration, and well-being of both migrants and the local population.
10. Investigate the role of technology in facilitating and shaping migration patterns and social dynamics. Examine how digital platforms, social media, and communication technologies influence migration decisions, information exchange, and community networking among migrants.

8

Limitations and Future Scope

8.1 Limitations

1. The research was limited to a specific geographic area.
2. The study was constrained by limitations in time and re*Source*s.
3. The study relies on the subjective opinions of the participants, which may be influenced by their individual perspectives and biases, despite the researcher's efforts to gather them as objectively as possible.
4. It would have been advantageous to include a broader range of relevant variables in the study.
5. A suitable statistical model could have been employed to provide thematic support for the behavior and functioning of the system.
6. The selection of variables could have been expanded by incorporating additional variables that would have demonstrated resilience in domain interactions.

8.2 Future Scope of Research

1. A longitudinal study can be conducted to examine the dynamics of migration in Darjeeling over an extended period. This would help capture changes in the social ecology of migration and identify emerging patterns and trends.
2. Comparatively analyzing the social ecology of migration in Darjeeling with other regions or similar contexts. This would provide a broader understanding of the factors influencing migration and their specific implications for the local social ecology.
3. Investigate the socioeconomic factors that contribute to migration in Darjeeling. Explore how economic conditions, employment opportunities, and income disparities affect the decision to migrate and shape the social dynamics within migrant communities.
4. Explore the mental models and perceptions of migrants regarding their chosen destinations. Investigate how these mental models influence decision-making processes, adaptation strategies, and social interactions within migrant communities.
5. Examine the social consequences of migration in Darjeeling. Investigate how migration affects social networks, community cohesion, cultural

practices, and identity formation, both among migrants and in the host community.

6. Assess the implications of migration on urban planning, infrastructure development, and public services in destination areas. Identify strategies to address the challenges posed by migration, including housing, transportation, healthcare, and education.
7. Investigate the relationship between climate change and migration in Darjeeling. Examine how environmental factors, such as changing weather patterns and natural disasters, interact with socioeconomic conditions to influence migration decisions and shape the social ecology.
8. Explore approaches to enhance community engagement and empower migrants in decision-making processes. Investigate how participatory methods and community-driven initiatives can promote inclusive development and address the social and economic needs of migrant populations.
9. Develop evidence-based policy recommendations to address the social ecological challenges associated with migration in Darjeeling. Provide insights on fostering sustainable development, social integration, and well-being of both migrants and the local population.
10. Investigate the role of technology in facilitating and shaping migration patterns and social dynamics. Examine how digital platforms, social media, and communication technologies influence migration decisions, information exchange, and community networking among migrants.

Bibliography

A Baddeley. (2003).Working memory: Looking back and looking forward. Nat Rev Neurosis 4: 829–839.

Acharya SK & Kaibartya D. D. (2022). Social Ecology of Migration: Reality and Realization. Astral International Pvt. Ltd, New Delhi, ISBN: ISBN: 978-93-5461-325-8 (HB)

Agesa, Richard U. and Sunwoong Kim (2001): "Rural to Urban Migration as a Household Decision: Evidence from Kenya". Rev. Dev. Econ. 5 (1): 60-75.

Almeida, C. C. (1973). Emigration, Espace et Sous-Dévéloppement. Int. Mig. XI (3):112–117.

Arango, J. (2000). Explaining migration: A Critical View. Int. Soc. Sci. J. 52(165): pp. 283-296.

Atma Ram & Sons

Atma Ram & Sons

Atma Ram & Sons.

Atma Ram & Sons.

Atma Ram & Sons.

Atma Ram & Sons.

Atma Ram & Sons.

Atma Ram & Sons.

Awasthi, I., and Mehta, B. S. (2020). Forced out-migration from hill regions and return migration during the pandemic: evidence from Uttarakhand. Indian J. Labour Econ. 63:1107-1124.

Barnum, H. N., & Sabot, R. H. (1977). Education, employment probabilities and rural-urban migration in Tanzania (Vol. 39, No. 2). Publication Unit, World Bank.

Barry, Roger G. (2008). Mountain Weather and Climate (3rd ed.). Cambridge University Press. ISBN 978-0-521-86295-0.

Bartel, Ann P. (1979): "The Migration Decision: What Role Does Job Mobility Play?"Am Econ Rev.69(5): pp: 775-786.

Bebbington, A. (1999). Capitals and Capabilities: A Framework for Analyzing Peasant Viability, Rural Livelihoods and Poverty. World Dev.27(12):2021–2044.

Beijer, G. (1970). International and national migratory movements. Int Migr. 8(3), 93-109.

Bernbaum, Edwin (2022). Sacred Mountains of the World (2nd ed.). Cambridge University Press. ISBN 978-1-108-83474-2.

Berry, J. W. (1997). Immigration, acculturation, and adaptation. Applied Psychology: An Interna-

Berry, J. W. (1997). Immigration, acculturation, and adaptation. Applied Psychology: An Interna-

Berry, J. W. (1997). Immigration, acculturation, and adaptation. Applied Psychology: An Interna-

Berry, J. W. (1997). Immigration, acculturation, and adaptation. Applied Psychology: An Interna-

Berry J.W. (1997). Immigration, acculturation, and adaptation. Applied Psychology:Int Rev.46(1), 5-34.doi:10.1111/j.1464-0597.1997.tb01087.x

Bolaria, B. Singh and Bolaria, Rosemary von Elling (1997). International Labour Migration. Oxford University Press, P- 1.

Breman, J. (1985). Of Peasants. Migrants and Paupers: Rural Labour Circulation and Capitalist Production in West India, Oxford University Press, New Delhi.

Brown, T., Ganguly-Scrase, R., & Scrase, T. J. (2016). Urbanization, rural mobility, and new class relations in Darjeeling, India. Crit. Asian Stud. 48(2):235-256.

Carney, D. E., ed. (1998). Sustainable Rural Livelihoods. What Contribution Can We Make? Papers presented at the DFID Natural ReSources Advisers' Conference. July 1998 London: DFID.

Castles, S. and Miller M. J.(2009). The Age of Migration. Houndmills, Basingstoke, Hampshire and London: MacMillan Press Ltd.

Chattopadhyaya, H. (1987). International Migration in India:. A Case Study of Bengal. K. P. Bagchi and Company. P -1-4, 6, 56-57,62-63, 377

Collinson, S. (2009). The political economy of migration processes: An agenda for migration research and analysis.. International Migration Institute,12.

Craik, K. J. W. (1943). The nature of explanation. Cambridge: Cambridge University Press.

Crawford, T. (1973): Beliefs about Birth Control: A Consistency Theory Analysis. Representative Research in Social Psychology. 4:pp. 53-65.

Curran, S. R., Shafer, S., Donato, K. M., & Garip, F. (2006). Mapping gender and migration in sociological scholarship: Is it segregation or integration?. Int Mig Rev.40(1):199-223.

Datta, A.(2003): Human Migration - A Social Phenomenon. Mittal Publications, P -43-44, 47, 51, 55, 57, 61, 66, 76, 78, 80-82, 87, 89-90, 160-161.

De Haan, A. (2011). Inclusive growth?: Labour migration and poverty in India. Netherlands: International Institute of Social Studies.

De Haan. A. (1999). Livelihoods and Poverty: The Role of Migration– A Critical Review of the Migration Literature. J. Dev. Stud. 36(2): 1-47

De Haas, H. (2005). International Migration, Remittances and Development: Myths and facts. Third World Quarterly 26(8):1269–1284.

De Haas, H. (2010b). Migration and development: A theoretical perspective1. Int. Mig. Rev. 44(1): pp. 227-264.

De Haas H.(2010): A Theoretical Perspective. Int. Migr. Rev. Mar;44(1):227-264. doi: 10.1111/j.1747-7379.2009.00804.x.

De Jong, G. F., & Fawcett, J. T. (1981). Motivations for migration: An assessment and a value-expectancy research model. In Migration decision making (pp. 13-58). Pergamon.

Djajic, S. (1986). International Migration, Remittances and Welfare in a Dependent Economy. J. Dev. Econ. 21:229–234.

Drew, G. R., & Rai, R. (2018). Connection amidst disconnection: water struggles, social structures, and geographies of exclusion in Darjeeling. Oxford University Press.

Dustmann, C., & Glitz, A. C. E. (2005). Immigration, jobs and wages: Theory, evidence and opinion. Centre for Economic Policy Research.

Ellis, F. (1998). Household Strategies and Rural Livelihood Diversification. J. Dev. Stud. 35(1):1–38.

Ellis, F. (2000). Rural Livelihoods and Diversity in Developing Countries. Oxford: Oxford University Press.

Everett S. Lee,(1966). A Theory of Migration. Demogr.3:pp. 47-57.

Faist, T. 2004. The Volume and Dynamics of International Migration and Transnational Social Spaces. Oxford: Oxford University Press.

Fei, J. C. H., & Ranis, G. (1964). Development of the labor surplus economy: theory and policy. Development of the labor surplus economy: theory and policy.

Funkhouser, E. (1995). 'Remittances from International Migration: A Comparison of Elsalvador and Nicaragua', Rev Econ Stat. 77(I), pp. 137-146.

Ganguly-Scrase, R., & Scrase, T. J. (2015). Darjeeling re-made: The cultural politics of charm and heritage. South Asia: J. South Asian Stud. 38(2):246-262.

Gerrard, John (1990). Mountain environments: an examination of the physical geography of mountains. MIT Press. ISBN 978-0-262-07128-4. OCLC 20637538.

Glick Schiller, N., Basch L. and Blanc-Szanton C. 1991. Towards a Transnational Perspective on Migration. New York: New York Academy of Sciences.

Glytsos, N.P. (1988). Remittances in Temporary Migration: A Theoretical Model and its Testing with the Greek - German experience, Weltwirtsch Arch. 124(4), pp. 524-549.

Gray, S. A., Gray, S., Cox, L. J., & Henly-Shepard, S. (2013). Mental modeler: a fuzzy-logic cognitive mapping modeling tool for adaptive environmental management. In 2013 46th Hawaii International Conference on System Sciences (pp. 965-973). IEEE.

Guarnizo, L., Portes A. and Haller W. (2003). Assimilation and Transnationalism: Determinants of Transnational Political Action Among Contemporary Migrants. Am.J.Sociol. 108(6):1211–1248.

Gugler, J. (1968). The impact of labour migration on society and economy in sub-Saharan Africa: Empirical findings and theoretical considerations. African Social Research, 6: 463-486.

Haan, A. D., Brock, K., Carswell, G., Coulibaly, N., Seba, H., & Toufique, K. A. (2000). Migration and livelihoods: case studies in Bangladesh, Ethiopia and Mali (No. 46).

Harbison, S. F. (1981). Family structure and family strategy in migration decision making. Migration decision making, 225-251.

Hossain, Md. Zakir (2001). 'Rural-Urban Migration in Bangladesh: A MicroLevel Study', Seminar Proceedings on Internal Migration, Brazil IUSSP Conference, August 20-24, 2001.

Johnson-Laird, P. N. (2010). Mental models and human reasoning. Proc. Natl. Acad. Sci. 107(43):18243-18250.

Johnson-Laird, P. N. and Byrne, R. M. (2002). Conditionals: a theory of meaning, pragmatics, and inference. Psychol Rev.109(4): 646.

Jokisch, B. D. (2002). Migration and agricultural change: The case of smallholder agriculture in highland Ecuador. Hum. Ecol. 30: 523-550.

Jong, G. D.and Fawcett, J. (1981). Motivations for Migration: An Assessment and a Value-Expectancy Model. De Jong and RW Gardner (eds), New York: Pergamon, 13-58.

Keely, C. and Tran B. N. 1989. Remittances from Labor Migration: Evaluations, Performance, and Implications. Int. Migr. Rev.23(3):500–525.

Khan, N. (1983). Studies in Human Migration. Rajesh Publications. New Delhi. P- 1, 5

Khemlani, S., Mackiewicz, R., Bucciarelli, M., and Johnson-Laird, P. N. (2013). Kinematic mental simulations in abduction and deduction. Proc. Natl. Acad. Sci. U.S.A. 110, 16766–16771. doi: 10.1073/pnas.1316275110

Khemlani, S., Orenes, I., and Johnson-Laird, P. N. (2012). Negation: a theory of its meaning, representation, and use. J. Cogn. Psychol. 24, 541–559. doi: 10.1080/20445911.2012.660913

Kindleberger, C. P. (1965). Europe's Postwar Growth: The Role of Labor Supply. New York: Oxford University Press.

Klein, Alexander (2011) "Did Children's Education Matter? Family Migration as a Mechanism of Human Capital Investment: Evidence from Nineteenth-Century Bohemia."Econ.Hist. Rev.64(3): 730–64.

Kroeber, A. L. (1948). Anthropology: Race. Language, Culture, Psychology, Prehistory, 856.

Kuznets, S. (1966). Modern Economic Growth, New Haven, CT: Yale University Press.

L. Markóczy and J. Goldberg (1995). A Method for eliciting and comparing causal maps. J Manage. 21 (2): pp. 305-333.

Lama.Mahendra .P (1986). 'Tea Industry and its Impact on Labour', Tea Plantation Workers in the Eastern Himalayas - A Study on Wages, Employment and Living Standards. Eds. R L Sarkar and Mahendra P Lama. Atma Ram & Sons. Delhi. Lucknow.P -20.

Lieten, G. K. and Nieuwenhuys O. (1989) Introduction: Survival and Emancipation" In Women, Migrants and Tribals: Survival Strategies in Asia. Ed. Lieten G. K., Nieuwenhuys O. and Schenk-Sandbergen L. New Delhi: Manohar, Pp. 1–8.

Lucas, R. E. B. and Stark O. (1985). Motivations to Remit: Evidence from Botswana. J. Pol. Econ. 93:901–918.

Mandi, Swati S. (2016). Natural UV Radiation in Enhancing Survival Value and Quality of Plants. Springer Nature. pp. 9–10. ISBN 978-81-322-2765-6.

Martin Schaffernicht, Stefan N. Groesser (2011). A Comprehensive method for comparing mental models of dynamic systems. Eur. J. Oper. Res. 210 (1): Pp. 57-67.

Mashqura, Fareedi and Lepcha. Pasang D. (2004). Area and Issue Profile of Darjeeling, Darjeeling Ladenla Road Prema R. C. D. C. Hayden Hall. P- 45.

Massey, D. S., Arango, J., Hugo, G., Kouaouci, A., & Pellegrino, A. (1999). Worlds in motion: understanding international migration at the end of the millennium: understanding international migration at the end of the millennium. Clarendon Press.

Massey, D. S., Arango, J., Hugo, G., Kouaouci, A., Pellegrino, A., & Taylor, J. E. (1993). Theories of international migration: a review and appraisal. Popul Dev Rev.19 (3): pp. 431-466.

McCarthy, N., Carletto, C., Davis, B., & Maltsoglou, I. (2006). Assessing the impact of massive out-migration on agriculture.

McCatty, M. (2004). The process of rural-urban migration in developing countries. An honours essay submitted in fulfilment of the degree of Bachelor of Arts to Department of Economics, Carleton University, Ottawa, Ontario.

McDowell, C. and De Haan A. (1997). Migration and Sustainable Livelihoods: A Critical Review of the Literature. Sussex: Institute of Development Studies.

Mehta, G. S..(1991).Socio-Economic Aspects of Migration. Deep and Deep Publications. p- 1, 3-4

Michael, P. (1979). Birds of Passage. Migrant Labour and Industrial Societies. Cambridge: Cambridge University Press.

Mincer, J. (1978). Family migration decisions. J.Polit Econ. 86(5):749-773.

Misra, Sib Ranjan (1986). Tea Industry in India. Ashish Publishing House. New Delhi. P-5-9.

Mondal, T. K., & Roychowdhury, P. (2019). Water scarcity in Himalayan hill town: a study of Darjeeling municipality, India. Urban drought: Emerging water challenges in Asia, 363-383.

Mosse, D., Grillo, R. D., & Stirrat, R. L. (1997). Seasonal labour migration in tribal (Bhil) Western India'(mimeo). Centre for Development Studies, University of Wales. UK.

Mosse, D., S. Gupta, M. Mehta, V. Shah, J. Rees and KRIBP Team (2002). 'Brokered livelihoods: Debt, labour migration and development in tribal western India', J. Dev. Stud. 38(5): pp. 59–88.

Myrdal, G. (1957). Rich Lands and Poor. New York. Harper and Row.

Papola, T.S. (1981). 'Urban Informal Sector in a Developing Economy'. Vikas Publishing House. New Delhi.

Peirce, C. S. (1931). 1958. Collected Papers of Charles Sanders Peirce, 8:1-6.

Peirce, C. S. (1931–1958). Collected Papers of Charles Sanders Peirce. Cambridge, MA: Harvard University Press.

Piper N (2009). The Gendered Political Economy of Migration - International Migration and Diaspora Studies Project - lib.jnu.ac.in.

PN Johnson-Laird, F Savary (1999). Illusory inferences: A novel class of erroneous deductions. Cognition .71:191–229.

PN Johnson-Laird, P Legrenzi, V Girotto, MS Legrenzi, J-P Caverni (1999). Naive probability: a mental model theory of extensional reasoning. Psychol Rev. 106:62–88.

Prakash, N. (2009). The development impact of workers' remittances in Fiji. http://mro.massey.ac.nz/xmlui/handle/10179/1281.

Rahamtulla, M., Roy, A. K., & Khasim, S. M. (2020). Orchid diversity in Darjeeling Himalaya, India: present status and conservation. Orchid Biology: Recent Trends & Challenges, 155-188.

Ritchey, P. N. (1976). Explanations of migration. Annu. Rev. Social. 2(1):363-404.

Roy, H. (1972). Some observations on new international tea agreement. Econ. 17(8):386.

Roy, N. C., & Biswas, D. (2021). Human ReSource Management in the Indian Tea Industry. Routledge.

Sandell, S. H. (1977). Women and the economics of family migration. Rev. Econ. Stat.59: 406-414.

Sarkar, Bjdyut (1984). Tea in India, Published by consultative committee of Plantation Associations. Calcutta. P- 1-2, 36-37, 41, 43-44.

Sarkar, F.L. and M.P. Lama (1986). Tea Plantation Workers in the Eastern Himalayas: A Study on Wages, Employment and Living Standards. Delhi. Atma Ram & Sons.

Sarkar, F.L. and M.P. Lama, 1986, Tea Plantation

Sarkar, F.L. and M.P. Lama, 1986, Tea Plantation

Sarkar, F.L. and M.P. Lama, 1986, Tea Plantation

Sarkar, F.L. and M.P. Lama, 1986, Tea Plantation

Sarkar, F.L. and M.P. Lama, 1986, Tea Plantation

Sarkar, F.L. and M.P. Lama, 1986, Tea Plantation

Sarkar, F.L. and M.P. Lama, 1986, Tea Plantation

Sarkar, F.L. and M.P. Lama, 1986, Tea Plantation

Sarkar, R. L. and Lama, Mahendra P. (1986).Tea Plantation Workers in the Eastern Himalayas - A Study on Wages, Employment and Living Standards. Atma Ram & Sons. Delhi. Lucknow: P- 2-3,5.

Schultz, T.N. (1971). 'Rural-Urban Migration in Columbia', Rev. Econ. Stat. 53(2):pp. 157-163.

Singh T. (1992). Tourism Environment: Nature Culture Economy. InterIndia Publications.New Delhi.

Sjaastad, L.A. (1962), The Costs and Returns of Human Migration. J. Polit. Econ. 70: 80-93.

Skeldon, R. (2008). Migration and development. In United Nations expert group meeting on international migration and development in Asia and the Pacific.

Snel, E.,Engbersen G. and Leerkes A. 2006. Transnational Involvement and Social Integration. Glob. Netw.6(3):285–308.

Srivastava, R. S. (1987). Surplus appropriation and accumulation by rural households in India: A case study based on fieldwork in Uttar Pradesh (Doctoral dissertation, University of Cambridge).

Stark, O. (1978). Economic-demographic interactions in agricultural development: The case of rural-to-urban migration (Vol. 6). FAO.

Stark, O. (1980). On the Role of Urban-to-Rural Remittances in Rural Development. J. Dev. Stud.16:369–374.

Stark, O. (1991). The Migration of Labor. Oxford: Basil Blackwell Ltd.

Stark, O. and Bloom D. E. (1985). The New Economics of Labor Migration. Am. Econ. Rev. 75:173–178.

Stark, O. and Levhari D. (1982). On Migration and Risk in LDCs. Econ Dev Cult Change. 31(1):191–196.

Taylor, J. E. (1986). Differential migration, networks, information and risk. Migration, human capital and development. 4:147-171.

Taylor, J. E.(1999). The New Economics of Labour Migration and the Role of Remittances in the Migration Process. Int Migr.37(1):63–88.

Taylor, J. E., and Wyatt, T. J. (1996). The shadow value of migrant remittances, income and inequality in a household-farm economy. J. Dev. Stud. 32(6):899-912.

Thompson, Richard H. (1996). Assimilation. In Encyclopedia of Social and Cultural Anthropology, vol. 1. David Levinson and Melvin Embe r, eds. Pp.112-15. New York: Henry Holt and Company.

tional Review ,46(1), 5–34. doi:10.1111/j.1464-0597.1997.tb01087.x

tional Review ,46(1), 5–34. doi:10.1111/j.1464-0597.1997.tb01087.x

tional Review ,46(1), 5–34. doi:10.1111/j.1464-0597.1997.tb01087.x

tional Review,46(1), 5–34. doi:10.1111/j.1464-0597.1997.tb01087.x

Todaro, M. P. (1969). A Model of Labor Migration and Urban Unemployment in Less-Developed Countries. Am Econ.Rev.59:138–148.

Trent Brown (2015) .Youth Mobilities and Rural–Urban Tensions in Darjeeling, India. South Asia: J. South Asian Stud.38(2): 263-275

Van Naerssen, T., Spaan, E., and Zoomers, A. (2008). Globalization, Migration, and Development. Global Migr Dev. 15(1)

Vertovec, S. (2002, February 14-15). Transnational Networks and Skilled Labour Migration: Paper presented at the Ladenburger Diskurs "Migration" Gottlieb Daimler- und Karl Benz-Stiftung, Ladenburg. Retrieved from http://libguides.murdoch.edu.au/APA/conference.

Vertovec, S. 1999. Conceiving and Researching Transnationalism. Ethn Racial Stud. 22(2):445–462.

Wages, Employment and Living Standards, Delhi:

Wages, Employment and Living Standards, Delhi:

Wages, Employment and Living Standards, Delhi:

Wages, Employment and Living Standards, Delhi:

Wages, Employment and Living Standards, Delhi:

Wages, Employment and Living Standards, Delhi:

Wages, Employment and Living Standards, Delhi:

Wages, Employment and Living Standards, Delhi:

Walter, M. (2002). Encyclopaedia of Population and Development (Vol - II): Population policy and future strategy. Sarup and Sons. New Delhi. P- 62, 67

Westbrook, L. (2006). Mental models: a theoretical overview and preliminary study. J. Inf. Sci. 32(6): 563–579.

Winthrop, R. H. (1991). Dictionary of concepts in cultural anthropology. Bloomsbury Publishing . USA.

Wolpert, J. (1965, December). Behavioral aspects of the decision to migrate. In Papers of the Regional Science Association (Vol. 15, No. 1, pp. 159-169). Berlin/Heidelberg: Springer-Verlag.

Wolpert, J. (1965, December). Behavioral aspects of the decision to migrate. In Papers of the Regional Science Association .15(1): pp. 159-169. Berlin/Heidelberg: Springer-Verlag.
Workers in the Eastern Himalayas: A Study on
Workers in the Eastern Himalayas: A Study on
Workers in the Eastern Himalayas: A Study on
Workers in the Eastern Himalayas: A Study on
Workers in the Eastern Himalayas: A Study on
Workers in the Eastern Himalayas: A Study on
Workers in the Eastern Himalayas: A Study on
Workers in the Eastern Himalayas: A Study on
Yap, L. Y. (1977). The attraction of cities: a review of the migration literature. J. Dev. Econ. 4(3):239-264.